# CAD Method for Industrial Assembly

# CAD Method for Industrial Assembly

Concurrent Design of Products, Equipment and Control Systems

Edited by

**A. Delchambre**
*Department of Applied Mechanics,
Université Libre de Bruxelles, Belgium*

JOHN WILEY & SONS
Chichester · New York · Brisbane · Toronto · Singapore

Copyright © 1996 by   John Wiley & Sons Ltd,
                      Baffins Lane, Chichester,
                      West Sussex PO19 1UD, England

                      National        01243 779777
                      International   (+44) 1243 779777

All rights reserved.

No part of this book may be reproduced by any means,
or transmitted, or translated into a machine language
without the written permission of the publisher.

*Other Wiley Editorial Offices*

John Wiley & Sons, Inc., 605 Third Avenue,
New York, NY 10158-0012, USA

Jacaranda Wiley Ltd, 33 Park Road, Milton,
Queensland 4064, Australia

John Wiley & Sons (Canada) Ltd, 22 Worcester Road,
Rexdale, Ontario M9W 1L1, Canada

John Wiley & Sons (SEA) Pte Ltd, 2 Clementi Loop #02-01,
Jin Xing Distripark, Singapore 0512

*Library of Congress Cataloging-in-Publication Data*

CAD method for industrial assembly : concurrent design of products,
  equipment, and control systems / edited by A. Delchambre.
     p.  cm.
  Includes bibliographical references and index.
  ISBN 0 471 96261 9 (hb : alk. paper)
  1. Assembly-line methods—Planning.  2. Computer-aided design.
  3. Concurrent engineering.  I. Delchambre, A.
  TS178.4.C33  1996
  670.42'7—dc20                                          95-48842
                                                             CIP

ISBN 0-471-96261-9

Typeset in 10/12pt Palatino by Acorn Bookwork, Salisbury, Wiltshire
Printed and bound in Great Britain by Bookcraft (Bath) Ltd
This book is printed on acid-free paper responsibly manufactured from sustainable
forestation, for which at least two trees are planted for each one used for paper
production.

# Contents

| | |
|---|---|
| Preface | ix |
| Acknowledgments | xv |
| Contributors | xvii |

**1 Introduction** — 1
 1.1 The Objective of the Book — 1
 1.2 The Context — 2
 1.3 The SCOPES Project — 3
 1.4 The Motivations — 3
 1.5 The Scenario — 4
 1.6 The Definition and Development Method — 5
 1.7 The CAD Method for Industrial Assembly — 8
 1.8 Organisation of the Book — 9

**2 The CAD Method for Industrial Assembly and Concurrent Engineering** — 13
 2.1 Introduction — 13
 2.2 Concurrent Engineering — 13
 2.3 Tools for Concurrent Engineering — 16
 2.4 An Integrated CAD Method for Concurrent Engineering — 18
 2.5 Conclusions — 24
 2.6 Bibliography — 25
 2.7 References — 25

**3 Proposed Architecture for the New CAD Method** — 27
 3.1 Introduction — 27
 3.2 The Three Main Flows of a Productive System — 29
 3.3 The Levels of Reality in the Design Flow — 37
 3.4 The Ideal and Chosen Architecture for a CAD Method — 44
 3.5 Conclusions — 48
 3.6 References — 49

| | | |
|---|---|---|
| **4** | **Product Design for Assembly** | **51** |
| | 4.1 Introduction | 51 |
| | 4.2 Users' Needs | 53 |
| | 4.3 State of the Art | 55 |
| | 4.4 Functionalities and Methodology | 57 |
| | 4.5 Conclusions | 92 |
| | 4.6 Bibliography | 92 |
| | 4.7 References | 93 |
| | | |
| **5** | **Assembly Planning** | **95** |
| | 5.1 Introduction | 95 |
| | 5.2 Users' Needs | 95 |
| | 5.3 State of the Art | 97 |
| | 5.4 Functionalities and Methodology | 101 |
| | 5.5 Conclusions | 125 |
| | 5.6 Bibliography | 125 |
| | 5.7 References | 126 |
| | | |
| **6** | **Resource Planning** | **129** |
| | 6.1 Introduction | 129 |
| | 6.2 Users' Needs | 132 |
| | 6.3 State of the Art | 133 |
| | 6.4 Functionalities and Methodology | 136 |
| | 6.5 Conclusions | 158 |
| | 6.6 References | 158 |
| | | |
| **7** | **The Simulation Module** | **161** |
| | 7.1 Introduction | 161 |
| | 7.2 Users' Needs | 163 |
| | 7.3 State of the Art | 166 |
| | 7.4 Functionalities and Methodology | 168 |
| | 7.5 Conclusions | 187 |
| | 7.6 References | 188 |
| | | |
| **8** | **The Scheduling Module** | **189** |
| | 8.1 Introduction | 189 |
| | 8.2 State of the Art | 190 |
| | 8.3 Users' Needs | 191 |
| | 8.4 Functionalities and Methodology | 193 |
| | 8.5 Conclusions | 212 |
| | 8.6 References | 214 |

## 9 The Flow Control Module — 215
9.1 Introduction — 215
9.2 Users' Needs — 218
9.3 State of the Art — 219
9.4 Functionalities and Methodology — 221
9.5 Conclusions — 235

## 10 Integration Aspects of the CAD Method — 237
10.1 Introduction — 237
10.2 Integrated Product Development Process — 238
10.3 Product Data Integration — 240
10.4 Man–Machine Interface — 250
10.5 The Man–Machine Interface in CATIA — 253
10.6 Conclusions — 262
10.7 Bibliography — 262

## 11 Introducing the Integrated CAD Method into Companies — 263
11.1 Introduction — 263
11.2 Organisation Impact — 264
11.3 Implementation — 266
11.4 Conclusions — 269
11.5 Bibliography — 269
11.6 Reference — 270

## 12 Conclusions — 271

## Index — 275

# *Preface*

**The Need for New Design Tools for Concurrent Engineering**

Concurrent Engineering (CE) is a promising approach to improve design time, cost, and quality. It aims to improve product development by combining, early in the design process, the constraints of product function, manufacturing, logistics, and many life-cycle issues. Those who have tried CE, however, agree that it is not an easy process to learn and sustain. Leaving aside institutional and interpersonal challenges, CE requires that an enormous amount of technical knowledge and tradeoffs be made explicit. These tradeoffs and knowledge were either ignored or handled implicitly in the past without documentation or systematic procedures.

To accomplish CE requires a new level of information structuring and integration, accompanied by new computer design tools and aids. Up to now, designers of complex electro-mechanical-optical (CEMO) products have not had a way to put any functional or non-geometric information into a CAD model while designing parts and products. Additionally, they have had no way to deal systematically with assemblies or sets of parts either geometrically or functionally. Instead they are limited to defining the geometry of individual parts and associating them with geometric constraints based on dimension lines or other drafting-based methods. No commercial CAD tools provide methods for creating groups of parts directly and linking them together in function-related ways.

By contrast, VLSI designers can access highly integrated tool sets that link concept design, basic modularity decisions, functional simulations, production tooling (i.e., photomasks), design, and process planning. VLSI designers can build complex designs by linking together part elements that they acquire from libraries complete with process plans and functional behaviour models. The designers can use these library elements with confidence because other engineers have pretested them, including their manufacturing processes.

No such set of tools exists for CEMO items; not only are most of the tools in the VLSI list missing, but what exists is not well integrated, except links between individual part geometry and fabrication methods. Furthermore, there is little standardisation of even elementary geometric items, even though there are many recognised advantages for doing so. An important reason for the lack of mechanical geometry libraries is the absence of a logically sound way to attach such elements to parts so that the advantages can be exploited. What is needed is not so much a method for creating such elements as a way of using them to link parts together, build product datamodels with them, and link them with manufacturing analysis and cost tools and product functional simulations.

## Assembly as the Integrator of the Product Design Process

It is likely that every design class (cars, aircraft, cameras, etc.) has its own set of technical drivers that determine how the design process should be organised. Discovering that process requires tools that are outside the scope of this book. Ulrich and Eppinger (Ulrich and Eppinger, 1995) describe product development methodologies in detail. Even if the design process for each type of product depends on its own technical factors, it is possible to identify some generic design processes that most CEMOs require and can be used to tie a design process together. Assembly is a most suitable generic process for this purpose.

Assembly has two features that make it especially important. First, it is inherently integrative, bringing together parts and therefore bringing together the designers and builders of those parts. Second, assembly is the moment when a product comes to life, inasmuch as single parts do not perform any functions by themselves. Thus assembly has a strong link to function. In many cases, product function follows the paths of assembly. In such cases, capturing the interconnection between parts is the first step in capturing product function. Researchers have recognised the ability of assembly—more generally, of "interfaces"—to formulate the basic factors and requirements of a design, capturing intent and relating it to geometry, function, and manufacturing. Keeping notes on interfaces is one way to build up a design history that keeps the main issues in focus.

## Manufacturing Innovation Based on Assembly

Forward-thinking companies have made assembly the platform for advanced manufacturing capabilities. Examples are Allen-Bradley (USA), Nippondenso (Japan), and Schneider Electric

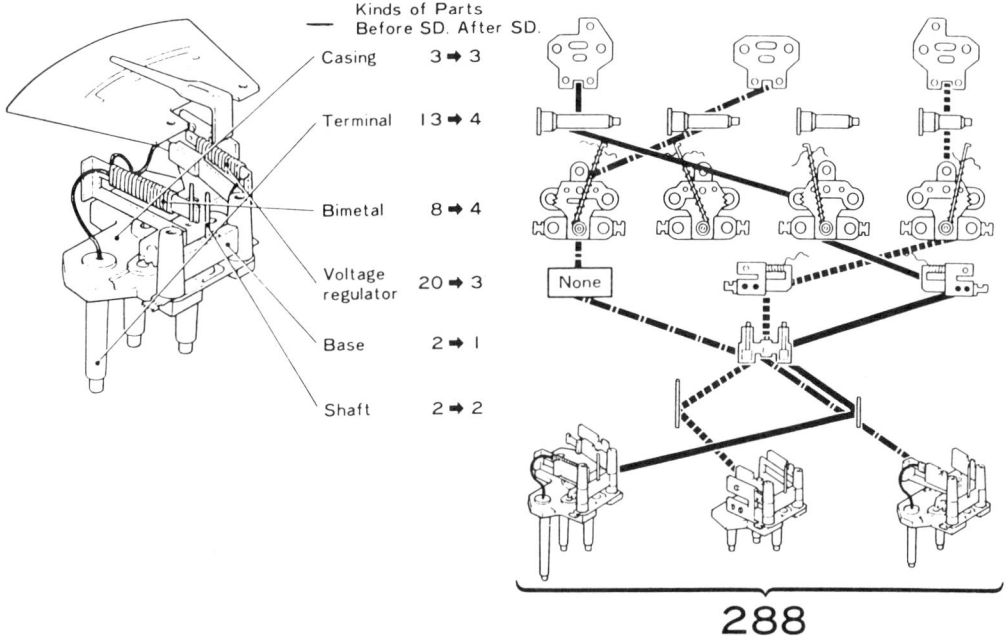

**Figure 1** Example of the Combinatoric Method of Model-Mix Assembly. Any one of 288 different panel meters can be assembled from the above set of 14 parts. Each path through the set at the right represents one valid model. These meters are assembled on an ordinary in-line fixed automation assembly machine with 3 to 4 part feeders at each station. A signal from the control computer indicates which part should be fed. For time efficiency, all items ordered of a given type are made in one solid batch and then the part feeders are switched over to permit assembly of a different type. Changeover takes only a few seconds.

(France). In each case, the company must manufacture CEMO products in unpredictable mixes, models, and quantities for a wide variety of customers. Each has carefully redesigned its products so that they can be assembled from a limited number of parts and subassemblies to meet a much larger range of end-products. In Nippondenso's case (Figure 1), a repertoire of only 14 parts can be assembled 6 at a time into as many as 288 different finished models.

Using assembly to differentiate product models instead of using part fabrication has many advantages. To follow this method, the designers may have to break a cardinal rule of Design for Assembly and create a larger number of simpler parts than DFA might indicate if its rules were applied to each product version alone. These simpler parts take hours or days to fabricate but only seconds to assemble. The method requires only that a sufficient stock of each part type be available. Deciding which to use takes a fraction of a second, and assembly takes little longer. Thus a new order can be responded to in much less than a minute, far less time than to fabricate the parts to order. This is called "Assembly-driven Manufacturing."

## Assembly in the Large—Assembly in the Small

Assembly can be regarded at two levels: the large and the small. Assembly in the small deals with the details of mating regions on parts and the physics of joining them. Examples include engineering models of screw thread mating or the assembly of pegs and holes. Tolerances, friction, the compliance of grippers, and speed of motion, among other factors, influence the reliability of assembly actions and the quality of assembly operations.

Assembly in the large deals with logical, logistical, financial, and operational issues of making products from parts. Topics to be dealt with include assembly sequence, incorporation of different versions of the product in series assembly, accomplishment of Just-in-Time operations, in-process testing, design for easy field repair, and so on. This book deals mainly with assembly in the large. A summary of assembly in the small may be found in Nevins and Whitney (1989).

## A Little History

A new phase in seeking to understand assembly began in the late 1960s with the advent of robots and the possibility of robot assembly. Robots have so little dexterity, sensing, or brainpower that assembly must be planned down to the last detail in order that robotisation will be successful. Early attempts to accomplish this merely revealed how little about assembly was really understood. Progress was made in the 1970s on robot programming, CAD models of parts and assembly constraints, machine vision, the physics of part mating, and other aspects of individual part assembly actions. Design for Assembly was also born during this decade. In general, the 1970s were the decade of assembly in the small.

In the 1980s, attention turned increasingly toward assembly in the large, along with the rise of CE. The pressure to combine functional and manufacturing constraints led to the recognition that much additional knowledge was needed to help designers bring CE into reality. This knowledge has been needed most at the system level of the design process, where the product's architecture is conceived and matching production systems are planned. Ironically, robot assembly played only a minor role in the 1980s, due to the lack of economical robot systems and the increased simplicity of products as a result of CE and DFA.

The 1980s also saw a great increase in the capabilities of computers as well as the software to support CAD for product design, particularly solid models. In the late 1980s, a number of these threads came together in the form of assembly sequence analysis and feature-based design applied to assembly modeling, (for example De Fazio et al., 1990). This work followed the tradition of earlier feature-based design, which had focused on joining CAD with metal cutting processes.

The present book represents an important step in continuing to rationalise design processes for CEMO products. It builds on the past and points the way to the future. It is the first to combine the off-line activities of product design for assembly and the on-line activities of scheduling and operating an assembly line. The work is the result of a multi-national combination of academic researchers and industrial companies who combined their knowledge and experience in several fields. It is to be hoped that the book and accompanying commercial software will find wide use in the education of new product and process designers as well as assisting experienced practitioners in industry.

*D. E. Whitney*
*Center for Technology, Policy, and Industrial Development, MIT*

## References

De Fazio, T. L., Abell, T. E., Amblard, G. P. and Whitney, D. E. (1990) Computer-Aided Assembly Sequence Editing and Choice: Editing Criteria, Bases, Rules and Techniques, *IEEE International Conference on Robotics and Automation*, pp. 416–422.

Nevins, J. L. and Whitney, D. E. (1989) *Concurrent Design of Products and Processes*, New York, McGraw-Hill.

Ulrich, K. T. and Eppinger, S. D. (1995) *Product Design and Development*, New York, McGraw-Hill.

# *Acknowledgments*

This book is the result of a fruitful European collaboration between two industrial companies: Schneider Electric (France) and Dassault Systèmes (France) and five Research Laboratories : Cranfield University (UK), CRIF and University of Brussels (Belgium), University of Stuttgart (Germany) and Ecole Polytechnique Fédérale de Lausanne (Switzerland).

Each laboratory was responsible for the writing of a given part of the book:

- University of Cranfield (Ip-Shing Fan and Gary Wallace) has mainly contributed to the Concurrent Engineering approach and the Design for Assembly tool (Chapters 2, 4, 10 and 11).
- CRIF and University of Brussels (Alain Delchambre, Emanuel Falkenauer and Alain Wafflard) have described the Assembly and Resource Planning modules (Chapters 1, 5, 6 and 12).
- University of Stuttgart (Rainer Heger) has reported the simulation environment (Chapter 7).
- Ecole Polytechnique Fédérale de Lausanne (Rafal Romanowicz and Eric Verdebout) have presented the architecture of the CAD method and the Scheduling and Flow Control modules (Chapters 3, 8 and 9).

My editing work was greatly simplified by the efficiency of all these contributors and I thank them very much for this.

The book is mainly based on the results of the ESPRIT III project 6562 SCOPES: *Systematic Concurrent Design of Products, Equipment and Control Systems*. Financial support was provided in part by the European Commission. Many thanks to the Commission of the European Union.

Finally, I would like to particularly thank Dr Daniel E. Whitney who has agreed to write the preface to this book. Dr Whitney has worked for more than 20 years and has made major contributions in the field of assembly process modeling and Concurrent Engineering, firstly, at the Charles Stark Draper Laboratory, Inc., and for the last two years at the Massachusetts Institute of Technology. Many thanks also to the research team who collaborated with him in this task: James L. Nevins, Thomas L. De Fazio, Samuel H. Drake, Alexander C. Edsall, Richard E. Gustavson, Anthony S. Kondoleon, Richard W. Metzinger, Jonathan M. Rourke, Donald S. Seltzer, Sergio N. Simunovic and Paul C. Watson.

# Contributors

| | |
|---|---|
| Ip-Shing Fan<br>Gary Wallace | The CIM Institute<br>Cranfield University<br>Cranfield, Bedford<br>MK43 0AL<br>UK |
| Alain Delchambre | Faculty of Applied Sciences<br>Department of Applied Mechanics<br>University of Brussels (ULB)<br>Avenue F.D. Roosevelt, 50<br>CP 165<br>B-1050 Brussels<br>Belgium |
| Emanuel Falkenauer<br>Alain Wafflard | Department of Industrial Management<br>and Automation<br>CRIF (Research Center for the Belgian<br>Metalworking Industry)<br>Avenue F.D. Roosevelt, 50<br>CP 106 – P4<br>B-1050 Brussels<br>Belgium |
| Rainer Heger | Institute for Human Factors and<br>Technology Management<br>University of Stuttgart<br>Nobelstrasse 12<br>D-70569 Stuttgart<br>Germany |
| Rafal Romanowicz<br>Eric Verdebout | Ecole Polytechnique Fédérale de Lausanne<br>Institut de Microtechnique<br>CH-1015 Lausanne<br>Switzerland |

# 1 Introduction

## 1.1 The Objective of the Book

The main objective of this book is to propose specifications and underlying concepts for future computer-aided tools to be used for the design and the control of flexible manufacturing systems for mechanical and electromechanical assemblies.

Such specifications and concepts will be applicable in the development of software tools covering product design at the upper level to shop-floor control at the lower one.

To reach this objective, the problem is approached under three main aspects:

- the "off-line aspect" is related to the product redesign at the upper level and to the "optimal" design of the equipment (workshop and cell layout, . . .) at the lower one.
- the "off-line and on-line part" deals with the discrete simulation which has to validate the equipment choices. This simulation takes into account scheduling and flow control aspects.
- the "on-line part" deals with the shop-floor control of the assembly (scheduling and flow control). The objective of concurrent design of equipment (using simulation) and shop-floor control is essential here and necessitated by the use of the same models and control functions within an off-line computer-aided design (CAD) system and on-line controllers.

The advances these specifications represent in relation to state-of-the-art processes and techniques are:

- The environment presented in this book is a real concurrent engineering one and it is the first time that the concurrent engineering steps are integrated in a CAD system,
- The "product redesign" is recommendation-oriented. In existing design for assembly (DFA) methods, little advice is

given on how to improve the design. As a redesign module will be integrated in a more general off-line workstation which includes assembly and resource planning, recommendations will also be generated by these two last modules. This integration allows the introduction of some more precise redesign rules in relation to the equipment than those implemented in the current DFA methods which are not linked to assembly and resource planners.
- The existing simulators do not take into account the rules of scheduling and flow control in the models of equipment. This point will be particularly addressed in this book by defining common models (equipments with their control) usable both in an off-line CAD system (for simulation) and in shop-floor controllers. This allows off-line and on-line validation of scheduling and flow-control rules.
- Usually, there is no connection between scheduling and flow control. In case of breakdown, the flow control result is changed according to this breakdown but regardless of the scheduling and of the current state of the lines and cells.

## 1.2 The Context

Competitive advantage is a result of being able to get better products to market faster than your competitors. After marketing has given the approval for developing a product, minimising the time to market or "beating the clock" is the number one objective of the product design team. The new working practice and tools adopted by companies to improve product development are known collectively as concurrent engineering.

Three objectives are usually pursued: to improve quality, to reduce lead time, and to reduce product cost. Each company may have a different balance of these objectives, but often improving one area will have beneficial effects on one of the other two.

The shift from serial prototype and analysis to concurrent engineering has enabled manufacturers to take a systematic approach to integrated product and process development. The birth of multidisciplinary teams is a result of implementing concurrent engineering practices and the need for employing the best resources for effective decision making. The capture, change and transfer of information becomes the main requirement for enabling concurrent engineering practices within the team. The requirements for information vary at different stages

# Introduction

of the product delivery process. Multidisciplinary teams in product and process development are seen as the key in increasing productivity and improving time to market.

Today, the concurrent engineering topic is a very crucial one, as it appears clearly that design and production offices can no longer be totally detached in a product's life cycle. Through the aid of an integrated computer system, designers of the product and designers of its related production means must be brought in front of a set of common data, understanding the consistency of their work from a product and production cost point of view, working in the best way so that the consequences of their design have as few drawbacks as possible in terms of impossibility to produce, high costs and low quality. This is particularly true in the domain of mechanical and electromechanical assemblies.

## 1.3 The SCOPES Project

The specifications and concepts presented in this book are mainly based on the results of the ESPRIT III project 6562: "Systematic COncurrent design of Products, Equipment and control Systems (SCOPES)". It is a three-year project, launched in July 1992, and regrouping a set of laboratories and industrials. The project partners are:

*Project coordinator*
Schneider Electric (France)

*Partners*

- Cranfield University (UK)
- CRIF—Research Center for the Belgian Metalworking Industry and University of Brussels (Belgium)
- Dassault Systèmes (France)
- EPFL—Swiss Federal Institute of Technology, Lausanne (Switzerland)
- University of Stuttgart (Germany)

## 1.4 The Motivations

The motivations for this project are the following ones:

- Assembly process represents, as an average, 40% of the product costs and 50% of the production investments are devoted to it.

- Design work's cost represents usually less than 10% of the total costs, but it freezes about 75% of subsequent costs through its important impact on the manufacturing processes.
- High investment costs of assembly automation can only be justified if the lifetime of the assembly system is longer than the product's one, meaning that flexibility is an absolute requirement. Perfect design for assembly and performance specification of product variants allow a wide-ranging application for an assembly system.
- The design for assembly, assembly planning and design of means of production can only be implemented, accepted and used in companies in an integrated software product.

## 1.5 The Scenario

The specifications and concepts defined and developed in this book are adapted to major assembly fields and related to user oriented requirements. One of the principal aims of the SCOPES project is indeed to answer real industrial worries and needs. Therefore, a complete study was made on the subject of product, process and equipment design in a concurrent engineering environment, by building case studies with the industrial partners and the members of the user group (this group is composed of industrial companies which have provided scenarios and case studies for enhanced practical solutions and specifications).

The application domain chosen to demonstrate the improvements in the concurrent engineering field is the assembly of mechanical and electromechanical products with the following characteristics:

- The cycle time of each assembly operation varies from 3 seconds to 1 minute.
- The annual production volume varies from 200 000 to 2 000 000 products per year.
- The production is medium-rated: from 200 to 2000 products per hour.
- The production is performed in a flexible way: production is done through variants of product batches.

The production means may be manual, semi-automatic or fully automatic with free or linked transfer systems.

Figures 1.1, 1.2 and 1.3 show typical pictures of these types of assembly lines.

**Figure 1.1** General view of an assembly line

## 1.6 The Definition and Development Method

The progress of the project was scheduled by exchanges between research institutes, users and Dassault Systèmes. Schematically, in the first part of the project—phase one—partners performed user's needs. The definition of the user's needs became the project reference. Every specification or development was judged from its ability to answer to the user's needs. Therefore, all partners were involved in the definition of this common reference.

The definition of the user's needs was done at the beginning by the research institutes. They conducted in parallel a gathering of a thorough state of the art in their respective fields. Then they defined the system that the project team had developed, broken down into potential functions. Practically speaking, these potential functions were proposed in a questionnaire submitted to the users. The questionnaire acted as a guideline to gather users' reactions.

In addition to the building of a common reference for project partners, users' needs analysis consisted in the definition of

**Figure 1.2** Presentation of assembly stations

case studies by users illustrating the test of the use of functionalities and verifying their abilities to answer the needs.

A user group was built to complete the set of applications. User group management was organised: regular meetings were convened to present latest prototype developments and results achieved with regards to the solution of their different case studies.

In a second phase, users' need analysis led to the definition of first specifications by the research institutes to initiate the developments of the demonstrations by Dassault Systèmes. The research institutes identified technical solutions to the user's needs and Dassault Systèmes integrated those solutions into prototypes, almost at the same time, due to time constraints. Technical solutions were described in terms of functional specifications, as a bridge between theoretical concepts and detailed specifications of the product.

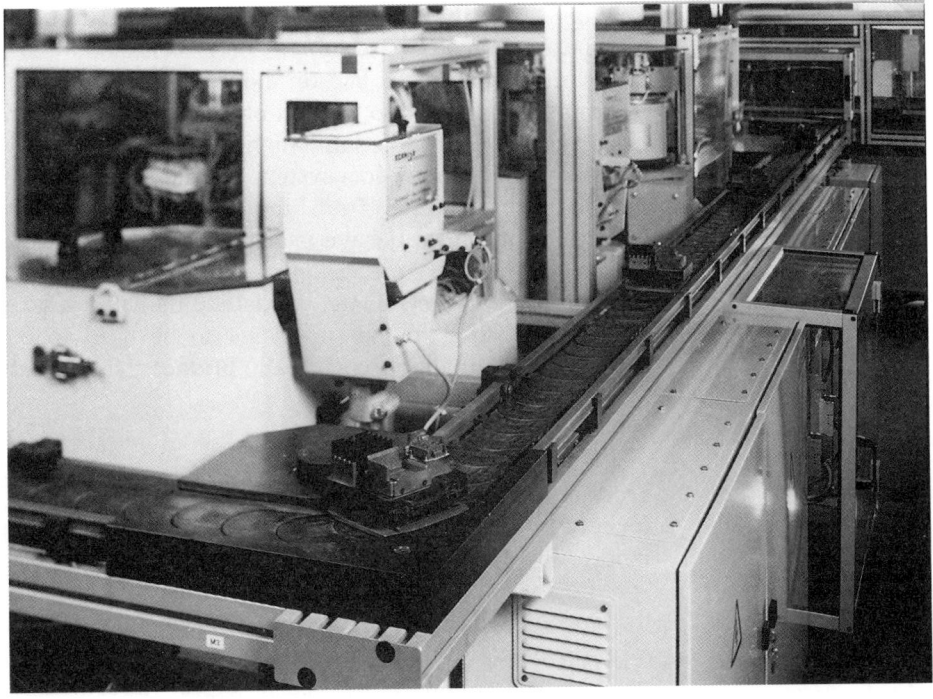

**Figure 1.3** Presentation of transfer system

In a third phase, users evaluated the specified functionalities with the help of demonstrations and gave a feedback to both the research institutes and Dassault Systèmes; which updated their work by providing enhanced specifications and including last results in demonstations before a final user evaluation. Evaluation was performed on the basis of the case studies. The use of prototypes rather than just documents was a main element in assessing the feasibility and usefulness of project results in terms of specifications and potential marketable products.

During the project, engineering of project results was organised. Also, partners studied the impact of their work at potential customer organisations through a study of the CAD system integration in a concurrent engineering environment. Finally, they spread project results, beside user group activities, with the organisation of user-oriented workshops.

## 1.7 The CAD Method for Industrial Assembly

The answer to the different needs is expressed through a CAD method for industrial assembly.

A strong emphasis is laid on the *dual approach*, consisting in a common treatment loop in both off-line and on-line systems. In the off-line system, the simulation module generates the shop-floor events, in the on-line system the monitoring module scans the shop-floor events from the line. The flow-control and scheduling modules behave exactly the same way in the CAD station and in the supervision system. In the CAD station, all the control parameters are validated, these parameters being later downloaded to the supervisor to be used to control the real line. This dual approach really bridges the gap between the off-line and the on-line worlds.

### 1.7.1 Off-line Modules

Product design is a totally advice-oriented design support function, without any automatic redesign of the product solid model. Its user-friendly interface allows the user to choose a specific analysis (should it be on the structure of the product, or the ease of handling, feeding, insertion or joining) for a part or a group of parts. Its architecture allows also the input of user-defined rules and other technological information on the product data structure.

The Assembly Planner constitutes a help to the designer who wants to represent in a clear and simple way, with user-friendly tools, an already evaluated assembly plan (formalism used to describe the assembly process), but can also be a complete tool for the generation of assembly constraints and the resulting assembly sequences. Manual creation of assembly plans, that reflects the know-how of the user, with standard icons and technological information capture capabilities, is also possible.

The Resource Planner allows a manual definition of the logical layout, with an integrated database for equipments. It also provides more automatic functions of line balancing and cost minimisation (obviously involving algorithms for the choice of equipements). The physical layout is also drawn in the module.

Then, this layout can be dynamically verified using the Simulation module. Simulation in the integrated CAD method for industrial assembly supplies the user with a simulated system in order to validate concurrently the solutions provided by the off- and on-line modules before implementing the real assembly system. Hence, the Simulation module will act as dynamic test bed for the on-line control software.

From the beginning of a design, the designer can call up any one of the modules to analyse the downstream functions and so modify the design to take into account its suggestions. The user can move between the various modules in any order, analysing the design against one or several considerations. In some cases, information is required from other modules, for example the resource planner requires the operations and precedence constraints generated from the assembly planner before an advanced optimisation of the layout. Early in the design process, much of the information needed by the modules will not be available. The information must be provided by the user, using intuition. As the design progresses and parts of the design become fixed and the level of uncertainty in the design reduces, more of this information can be extracted from the product model.

## 1.7.2 Common Modules

The common modules concern the control functions: scheduling and flow-control. The purpose of the scheduling module consists in defining the best possible sequence of production orders from a given list of customer demands, according to a strategy defined by the user. Then the flow control is responsible for the launching and the follow-up in real time of these production orders in the workshop. This module is event-driven: it has to take real-time decisions when events occur, following user strategy. In the off-line system, these events are created by the simulation module, in the on-line system the flow control answers to the shop-floor events. It must be added that the flow control in the on-line system is split into two levels: the global flow-control at the supervisor level and the local flow control at the PLC (programmable logic controller) level. During the design of the assembly line in the CAD station, the workshop management rules are set up: strategies are chosen. These management rules are then downloaded in the supervisor to handle the real assembly line. The downloading of control parameters ensures the usefulness of simulated results.

## 1.8 Organisation of the Book

This book is divided into twelve chapters.

The essence in concurrent engineering is seamless communication between the design, manufacture, production and all product- and process-related departments of a company. The core of the business activity is the product and its related information.

Chapter 2 shows how the CAD method for industrial assembly can facilitate the common access and analysis of the product and process definitions to support concurrent engineering.

Chapter 3 proposes an innovative architecture for the CAD method for industrial assembly which combines the different design activities in a concurrent engineering approach, instead of the traditional sequential approach.

The production of a product involves three main kinds of flows: design flow, material flow and control flow. During the concurrent design of the product and system, the design flow is highly emphasized and has generated the actual architecture.

This architecture is presented in more conceptual terms and related to other solutions in the litterature.

Product redesign for assembly is a new topic and must be well practised in a systematic and structured manner. The volume of knowledge in design for assembly is vast due to the diversity of manufacturing processes. For each industrial company, the specific nature of manufacturing plant and process capabilities reduces the application knowledge to a more manageable proportion. The redesign of a part is not an automatic process. The solid model of a part does not contain all the necessary data and knowledge for analysis. Assembly-oriented technological information on the product is maintained for analysis.

Chapter 4 presents a method to capture and maintain the manufacturing know-how and capability of the company. This product redesign analysis is a structured procedure assisting the user in identifying the critical issues in the product design and providing advice on possible redesign alternatives. The foci of analysis are the product structure and the ease of assembly.

Chapter 5 specifies the requirements for the Assembly planning module which are close to the product redesign ones. It must first provide tools to maintain, complete and enrich a database adapted to the know-how in deriving assembly plans, with some generic knowledge, and files corresponding to particular projects or products. Both geometrical and technological constraints must be taken into account to produce realistic assembly plans. The user-friendliness of the module is also very important with simple tools to produce, create or evaluate all possible assembly plans, and have a clear representation of them.

The Resource Planner is a tool to design both logical and physical layouts, taking into account their respective constraints (hard constraints pertaining to the choice of equipment and cycle time, and soft constraints concerning the optimisation of the planning process, with criteria like cost or reliability). It must be a help for the selection of equipment by providing a user-friendly and integrated database. Once again, the system must be open, so that user's knowledge can be easily stored and taken into account. One must be able to evaluate the logical layout solution quantitatively (global cost, maintenance requirement, and more), before constructing the physical layout by adding physical elements.

This subject is treated in Chapter 6.

Chapter 7 presents the simulation module which has become an essential tool in dealing with manufacturing and assembly system problems, in the design, implementation and operation stages. During the design phase, simulation assists the decision-making process to define features, kinds of equipment, layout, control, scheduling strategies and contingency plans when failures occur. Furthermore, simulation is a useful way to test the actual control software of the planned system before implementing the real process. During the system validation, the simulation module supplies the on-line modules with simulated system states. This allows the concurrent test of the flow-control and scheduling strategies. The simulation tool does not require model skills. Therefore, a strong emphasis is laid on the interface with the user in the simulation module during the build-up of the workshop model.

Chapter 8 describes the scheduling module. It is an automated scheduling aid (only proposes schedules) at the workshop level. It is divided into global scheduling, or scheduling of the workshop or cell, and local scheduling, or scheduling of a resource.

During the off-line phase, the module aids in the choice of the most adapted scheduling method. During the on-line phase, the module aids in real-time scheduling, and error recovery.

The flow control is responsible for the launching and the following in real time of the production orders (defined by the scheduling function) in the workshop. This module is event-driven: it has to take real-time decisions when events occur, following user strategy. These events are created by the simulation module in the off-line system. During the design of the assembly line in the CAD station, the workshop management rules are set up: strategies are chosen. These management rules are also downloaded in the supervisor which handles the real assembly line.

During the exploitation phase, the actual flow control is event driven as well but the events are transmitted through the hardware and network between the supervisor and the decentralised controllers.

Chapter 9 presents the highly innovative flow-control module which applies the most recent techniques in decentralised control.

The CAD method for industrial assembly is oriented towards concurrent engineering. Chapter 10 shows how, from the beginning of a design, the designer can call up any one of the modules to analyse the downstream functions and so modify the design to take account of its suggestions. The user can move between the various modules in any order, analysing the design against one or several considerations.

Concurrent engineering is a people and organisation approach. The success of any concurrent engineering implementation depends on the effective organisation and training of the work force. These organisation, training and implementation issues are introduced in Chapter 11.

Chapter 12 summarises the major contributions of this book.

# 2 The CAD Method for Industrial Assembly and Concurrent Engineering

## 2.1 Introduction

The success of world class companies during recent years can be attributed to the way they have managed the design process to produce manufactured goods of high quality, quickly and on time (Charney, 1991). This is due to technological innovation and by paying attention to the quality of the design. Other organisations have failed to realise until recently that the most effective way to eliminate cost is through a better concept design rather than organisational and technical efficiency.

It is now recognised that *design* means the process of developing for the market highly competitive, innovative, high added-value products that people want to buy. Concurrent engineering techniques have been implemented to encourage this philosophy. Computerised tools have been developed to enable a greater control over the engineering activities to encourage a correct first time design.

## 2.2 Concurrent Engineering

Competitive advantage is a result of being able to get better products to market faster than your competitors. After marketing has given the approval for developing a product, time to market is the predominant objective of the product design team. The new working practices and tools adopted by companies to improve product development are known collectively as concurrent engineering (CE) (Brophy et al., 1993). The objectives of concurrent engineering are generally agreed to be:

- Improving product quality—or the extent to which a product satisfies customer requirements. This has both subjective and objective attributes.
- Reducing lead time—that is, reducing the time from product concept to successfully bringing the product to the market (often termed "time to market").
- Reducing product cost—where product cost can be defined as the level of resources required to take the product from concept to market. This will include the hours worked on the product, materials used in the product, and any equipment or services that are used.

Each company may have a different balance of the three objectives, but often improving one area will have beneficial effects on the other two. The focus of concurrent engineering is to deliver cost-effective engineering and manufacture of a product (Fan, 1993).

The shift from serial prototype and analysis to concurrent engineering has enabled manufacturers to take a systematic approach to integrated product and process development. The birth of multi-disciplinary teams is a result of implementing concurrent engineering practices and the need for employing the best resources for effective decision making. The capture, change and transfer of information becomes the main requirement for enabling concurrent engineering practices within the team. The requirements for information vary at different stages of the product delivery process. Multi-disciplinary teams in product and process development are seen as the key in increasing productivity and improving time to market.

In 1991 the Concurrent Engineering Forum that took place at the CIM Institute at Cranfield University defined concurrent engineering as "the delivery of better, cheaper, faster products to market by a lean way of working using multi-discipline teams, right first time methods and parallel processing activities to continuously consider all constraints."

Industrial practice of concurrent engineering is still being developed. In 1990, the Eureka IDAP—Integrated Design and Assembly Planning project (Richter, 1991) used a questionnaire approach to complete a study on design and manufacture. Responses were obtained from 51 industrial companies. The design departments of less than 35% of the organisations cooperate with other departments in working groups. The design departments of 20% of the organisations talk to the production departments only in problem cases.

An extensive study of the global aerospace industry (Sehdev et al., 1995) found that industry realised that design for manufacture could bring significant benefits. However, the knowledge and experience of the individuals in the organisations is the main tool to practise design for manufacture. The industry is looking for suitable tools to support concurrent engineering.

The general area of design for manufacture and concurrent engineering are supported under the European Commission research and development programmes. The list of projects includes:

- **BRITE-EURAM**
    - Design methodologies for engineering component properties—PRECEPT.
    - Simultaneous engineering system for applications in mechanical engineering—SESAME.
    - The design, implementation and test of a design for manufacturing architecture and tool suite—DEFMAT.
    - Application of feature based modelling for complex product design and manufacture—FEMOD.
    - An intelligent knowledge assisted design environment—IKADE.
- **ESPRIT**
    - Computer-aided concurrent integral design—CACID.
    - Systematic concurrent design of products, equipment and control systems—SCOPES.
    - Concurrent and simultaneous engineering systems—CONSENS.
    - An integrated system for simultaneous bid preparation—BIDPREP.
    - Architecture, methodology and tools for computer-integrated large-scale engineering—ATLAS.
- **RACE**
    - Distributed industrial design and manufacturing of electronic sub-assemblies—DIDAMES.
    - An environment for distributed integrated design—EDID.

These projects cover a range of activities and represent the European view towards research and development in the design for manufacture and concurrent engineering areas.

## 2.3 Tools for Concurrent Engineering

Communication links between the team becomes a necessary requirement for enabling concurrent decision making. Besides the hardware, there exists a requirement for establishing a central database for collecting and distributing product system requirements. This database has to be designed so that each member of the team can access information in the format that is of most value to him or her.

The use of computer-supported cooperative work (CSCW) is relatively new. These new tools require the integration of existing CAD, CAM (computer-aided manufacturing) and CAE (computer-aided engineering) tools into the new environment. Sevenler et al. (1993) have highlighted requirements for a fully integrated CAD/CAM/CAE environment.

- Knowledge bases should be established for the retrieval and storing of corporate knowledge.
- There should be a central database to provide a flexible way of accessing, displaying, interpreting and distributing product information to all members of the team.
- The design teams should be provided with tools with a high degree of flexibility to promote creativity and quality engineering to optimise the subsystem design.
- Communication of data in various formats should be supported.
- The design applications and databases must provide a variety of evaluations and trade-off assessments. This allows iterations while still maintaining team consensus over global changes.
- The computer-supported negotiation tools that allow several users to access the models together with a common simulation environment and on-line graphical group interaction are needed to support concurrent engineering teams.
- The design system should enable the exchange of information between CAD, CAM and CAE, problem management and configuration control systems. Translations should be minimised to preserve the original data as much as possible.
- Effective data management should be performed to preserve the data relationships, capture the dependencies and the important characteristics of the design.
- There is a need for an interface to standard parts catalogues.
- All applications should be consistent and use the established design rules.
- Computer systems should support decision making rather than try to automate everything, since humans are still better in many areas than computer systems.

These can be summarised as follows:

- Communication between software modules: quick and easy transfer of information to any module.
- Operation of the module with only partial results: this means that the activities of a module can be started before the completion of the activity of another module (parallelisation) and therefore the interaction between modules can be really effective (early feedback).
- Data coherence between different modules: all modules should have common user interfaces (MMI standards), data structures and information semantics to avoid errors and inefficiencies.
- Operation of modules by non-experts: all designers must be able to use (albeit not in an optimised way) the modules where collaborations could influence their specifications, e.g. a mechanical designer must be able to use a planning module to obtain time and cost estimates.

The value of using tools is not necessarily in the improved figures they produce on paper or on a computer screen, but the new team culture that results and the impact this has on the product. An increase in individual awareness also results.

Concurrent engineering tools can be divided into two categories. Those that promote the consideration of manufacturing, quality, assembly, criteria throughout the design process and those which aid and improve the communication between team members.

As very comprehensive data volumes evolve they have to be efficiently communicated and expanded in concurrent engineering. Computerised models of product, processes and factory models are essential (Sohlenius, 1992).

There are many obstacles to a successful CE tool implementation. These include (Willcox, 1993):

- Inadequate Product Introduction Process (PIP) planning
- Lack of management commitment
- Strategic issues dominating
- Lack of resource early on
- Too strong focus on development
- Cost data in the wrong form
- Poor selection of tools

Concurrent engineering represents a change in the way of work of design and manufacturing personnel. Strong and committed management is needed to ensure that the organisation and technical barriers are removed (Fan, 1995).

One of the major technical barriers is the representation of product and process data. Design and manufacture departments have different requirements on their computing systems. This results in the use of a variety of computing tools in the company. The main obstacle then becomes the communication required between the different tools selected by different members in the project teams. The introduction of standards such as PDES/STEP may promise a solution, there are still development work in applications protocols before these exchange standards can be put into regular use.

As the main product characteristics are defined through geometry, a unifying platform based on a CAD system provides common access to an unambiguous single product description. The concurrent development of the product, assembly process and factory from the same set of data prevents any errors from mis-interpretation by different members in the concurrent engineering team.

Current advances in CAD technology provides satisfactory geometric representation of the product in solid models. Correspondingly advances in product modelling allows the capture of non-geometric and manufacturing information of the product with the necessary associative links to the geometric model. The Boeing 777 programme successfully used CAD technology to completely model the aircraft before manufacture (Gottschalk, 1994). An integrated CAD method to support Concurrent Engineering is now technically feasible.

## 2.4 An Integrated CAD Method for Concurrent Engineering

As mentioned in the previous chapter, a consortium of research laboratories and industrial companies in Europe completed the design of an integrated CAD method for concurrent engineering. It was supported by the European Commission under the Esprit programme. The SCOPES project (systematic concurrent design of products, equipment and control systems) completed the integration of product design to manufacture operations.

The integrated CAD method for concurrent engineering aims to:

- improve the communication between design and manufacturing members of the CE team, and
- provide a suite of integrated tools that provides design support on the downstream functions associated with the assembly of a product throughout the design process. These tools analyse the ease of assembly of a design, the assembly order, the design of an assembly workshop, the simulation of the workshop taking into account scheduling and flow control and control of the on-line workshop based on the results of the simulation.

A suite of prototype software modules has been produced by the industrial partners in the consortium to validate the concepts.

The off-line software modules have been developed by Dassault Systèmes, to run on the CATIA design system. The modules support the design of products, processes and factories before the physical construction. These off-line modules are:

- *Product design for assembly (DFA).* This provides positive advice to the user in how the design might be improved to ease its assembly throughout the design process.
- *Assembly planning (AP).* This generates an optimum assembly plan from geometrical and topological information about the product.
- *Resource planning (RP).* This helps the user to produce a logical layout, that is, regrouping the operations on cells or stations, attributing sets of equipment to each of them, while being constrained by a given cycle time. A physical layout editor enables the user to design the layout of the assembly workshops and cells from the logical layout.
- *Simulation.* used to validate the choice of the workshop and the cell architecture. The model of the workshop and cells will be applied in the on-line part for control functions.

The on-line supervision system was implemented by Schneider Electric to control automated assembly systems in real time. The modules include:

- *Scheduling and Flow Control.* Scheduling and flow control rules assign products and operations to the elements of the production system. Orders, scheduling and flow control represent the dynamic factors in the simulation model.

- *Shop-floor Control.* The resulting specifications achieved by the off-line system will be used for the effective shop-floor implementation and especially for the application development of the control system.
- *Monitoring.* In addition to a standard monitoring tool, the project focuses on a real integration of control functions.

These modules are integrated through the SCOPES architecture (see Figure 2.1).

The extent of integration of the process and factory design and control is a unique feature of the CAD method. Communication between design and manufacturing is improved as both members of the team are required to provide information to enable the analyses to be performed correctly.

For concurrent engineering to work effectively, it is necessary for the integrated CAD method to be able to approach the design of a product from two perspectives. The first is when the design is only represented as functional components, or perhaps as sub-assemblies. The second is when considering individual parts that combine to form a sub-assembly which in turn is just one part of the complete product (see Figure 2.2)

When considering either of these scenarios all the modules can be run from early within the design process to enable the user to consider as many of the downstream processes as possible (see Figure 2.3).

The integrated CAD method is flexible in the way that it

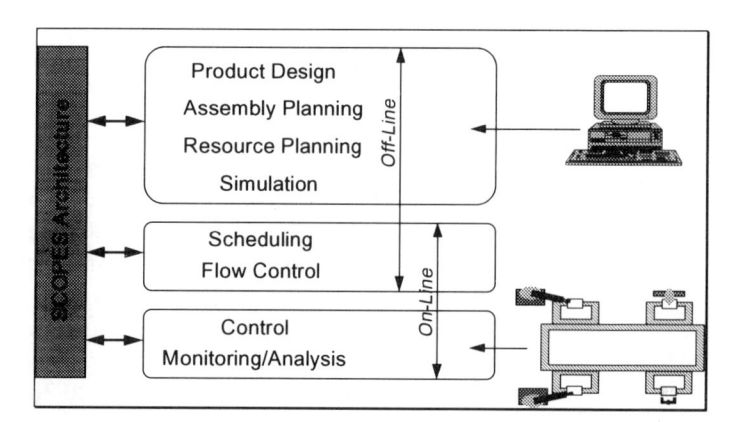

**Figure 2.1** Integrated CAD method for concurrent engineering

# The CAD Method for Industrial Assembly and Concurrent Engineering

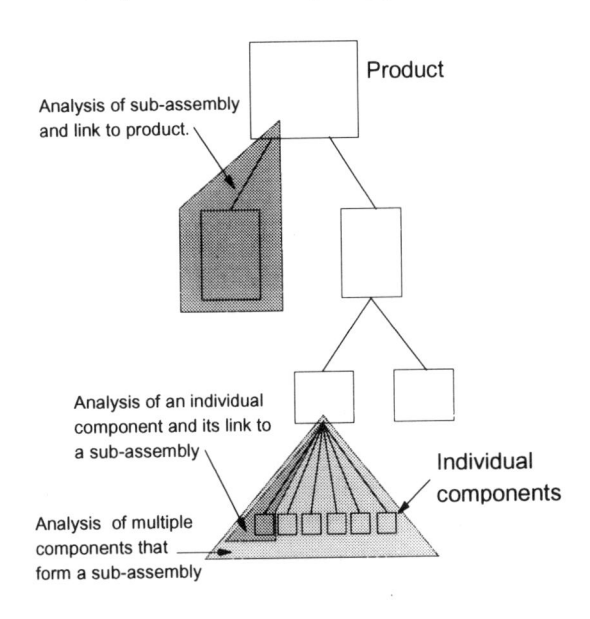

**Figure 2.2** Analysis of products and components

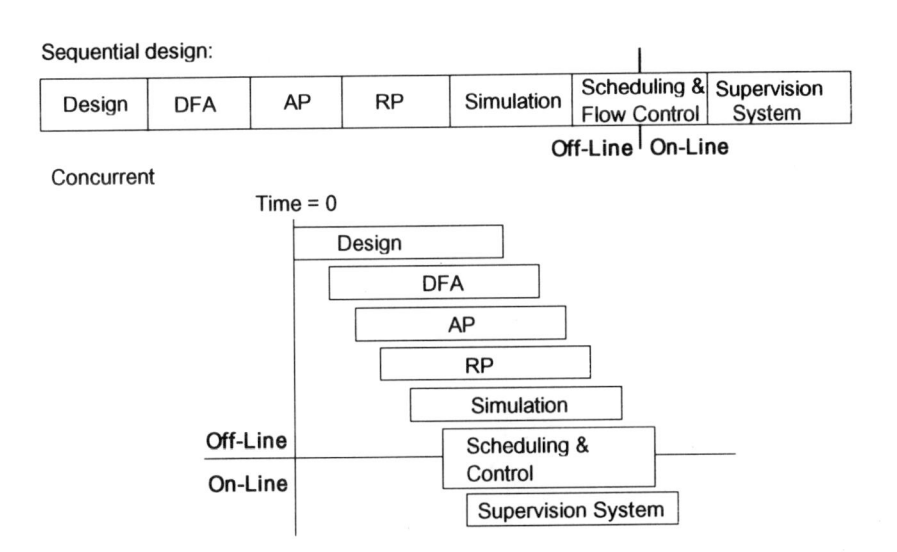

**Figure 2.3** Sequential and concurrent design

might be used. From the beginning of a design the designer can call up any one of the modules to analyse the downstream functions and so modify the design to take into account its suggestions. The user can move between the various modules in any order, analysing the design against one or several considerations. In some cases information is required from other modules, for example the Resource Planner requires the operations and precedence constraints generated by the Assembly Planner before making an equipment selection. In these cases the appropriate module would be run automatically or the user asked to provide an input. The Design For Assembly and Assembly Planning modules do not necessarily require the geometry of components before an analysis can begin. Early in the design process much of the information needed by the modules will not be available. This information must be provided by the user using his experience. As the design progresses and parts of the design become fixed and the level of uncertainty in the design decreases, more of this information can be extracted from the product model (see Figure 2.4).

This flexibility is illustrated in Figure 2.5 where free movement between the off-line modules is illustrated. This is of course possible between all the modules, both off-line and on-line.

A summary of the coherence of each analysis module with the concurrent engineering approach is illustrated below.

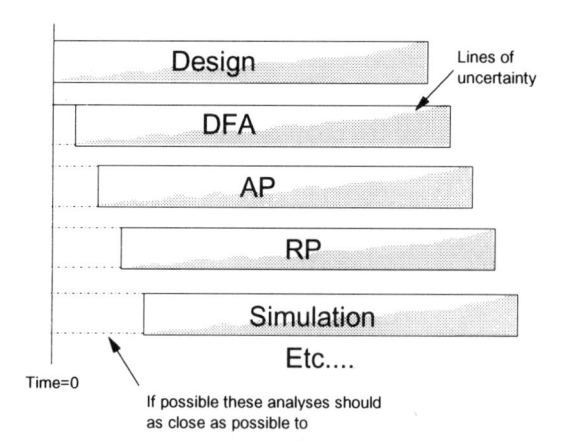

**Figure 2.4** Information certainty grows with design

# The CAD Method for Industrial Assembly and Concurrent Engineering

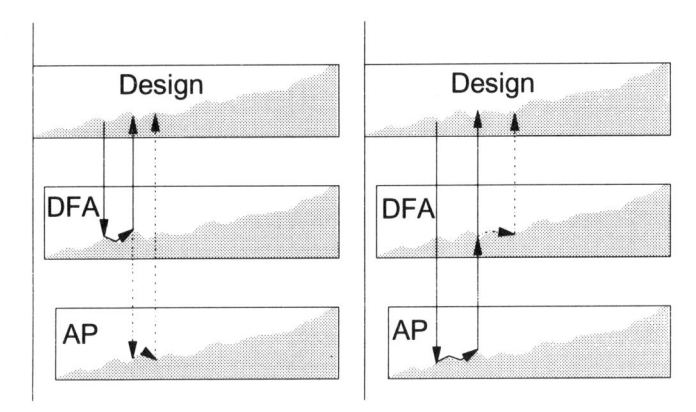

**Figure 2.5** Flexibility within integrated CAD method

### 2.4.1 Product Design For Assembly Module

The Product Design for Assembly (DFA) module supports the consideration of assembly factors at the conceptual stage of the product design process.

- It is directly linked with the Assembly planner and Resource planner through the common database.
- It is indirectly linked with the other modules through the integrated architecture.
- The module can be used at any stage of the product development.
- It has a coherent Man–Machine interface (MMI) through the CATIA environment.

### 2.4.2 Assembly Planner

- Linked directly with the other off-line modules through the database and indirectly with the other modules through the integrated architecture.
- It is possible to define a draft assembly plan without the geometry of assembly, in order to have preliminary time and cost estimates.
- The CATIA environment is used so the MMI is consistent.
- The module is generative and does not require an expert to generate an assembly sequence.

### 2.4.3 Resource Planner

- Linked directly with the other off-line modules through the database and indirectly with the other modules through the integrated architecture.

- Possible to obtain preliminary indications on resources from draft assembly plans.
- The CATIA environment is used so the MMI is consistent.
- It is a generative type module that is able to give acceptable results with minimum user interaction.

## 2.4.4 Simulation

- Linked directly with both the off-line and on-line modules through the integrated architecture.
- Possible to obtain preliminary indications of the scheduling and flow control rules selected.
- The CATIA environment is used so the MMI is consistent.
- It is a generative type module that can be used to create a preliminary plan by a user not particularly familiar with simulation.

## 2.4.5 Scheduling and Flow Control

- Linked directly with the on-line modules and indirectly with the off-line modules through the integrated architecture.
- Possible to obtain preliminary indications of the scheduling and flow control rules selected and demonstrate them through the simulation.
- The CATIA environment is used so the MMI is consistent.
- It is a generative type module that can be used to give acceptable results with minimum user interaction.

## 2.4.6 Shop Floor System

- Linked directly with the on-line modules and indirectly with the off-line modules through the integrated architecture.
- MMI is designed to be as close as possible to the off-line interface bearing in mind the limitations of the system.
- Uses the same flow control and scheduling rules as generated by the off-line modules.

## 2.5 Conclusions

It can be concluded that the integrated CAD method provides a suite of tools that operate in a concurrent way by providing:

- The free exchange of data between modules that is possible throughout the design process and through to the operation of the workshop.
- The ability to use any one of the tools from very early within the design process.
- The suite of tools are completely integrated.

## 2.6 Bibliography

*21st Century Manufacturing Enterprise Strategy*, 1991, Iacocca Institute, Lehigh University.

Busby, J. S., Fan, I.-S. (1993) The extended manufacturing enterprise: its nature and its needs, *Intl. J. Technology Management*, **8**(5–6).

Davenport, T. H. and Short, J. E. (1990) The new industrial engineering: information technology and business process redesign, *Sloan Management Review*, **31**(4).

Ettlie, J. E. and Stoll, H. W. (1990) *Managing the Design-Manufacturing Process*, McGraw-Hill, New York.

Hammer, M. (1990) Reengineering work: don't Automate, Obliterate, *Harvard Business Review*, **68**(4).

Nevins, J. L. and Whitney, D. E. (1989) *Concurrent Design of Products and Processes*, McGraw-Hill, New York.

Pahl, G. and Beitz, W. (1988) *Engineering Design*, Design Council.

Parsaei, H. R. and Sullivan, W. G. (eds) (1993) *Concurrent Engineering*, Chapman & Hall, London.

Pugh, S. (1990) *Total Design*, Addison-Wesley, Reading, MA.

Shingo, S. (1989) *A Study of the Toyota Production System from an Industrial Engineering Viewpoint*, Productivity Press.

Tomiyama, T. and Yoshikawa, H. (1987) *Extended General Design Theory, Design Theory for CAD*, Elsevier, Amsterdam.

Yoshikawa, H. (1981) General design theory and a CAD system, *Proc. IFIP WG5.2/5.3 Working Conference* (Tokyo), North-Holland, Amsterdam.

Womack, J. P. et al. (1990) *The Machine That Changed The World*, Rawson Associates, New York.

## 2.7 References

Brophy G., Lettice, F., Sackett, P. and Fan, I.-S. (1993) Concurrent Engineering with CATIA, *European Catia User Association Annual Conference*.

Charney, C. Y. (1991) *Reducing Product Lead Time*, SME Publishing, Michigan.

Fan, I.-S. (1993) Design for manufacture and assembly in concurrent engineering, *Proc. 2nd Intl. Conf. Manufacturing Technology*, 15–18 December, Hong Kong.

Fan, I.-S. (1995) *DFM/EE Report on Characteristics of Design for Manufacture in the Extended Enterprise and Current World Practice in the Aerospace Industry*, The CIM Institute, Cranfield University, January 1995.

Gottschalk, M. A. (1984) How Boeing got to 777th heaven, *Global Design News*, November/December.

Richter, M. (1991) Eureka-Famos Project EU 289 IDAP, *2nd FAMOS Advanced Course on Flexible Automated Assembly*, Venice, 6–12 October 1991.

Sehdev, K., Fan, I.-S., Cooper, S. and Williams, G. (1995) Perceptions of design for manufacture in the aerospace extended enterprise, *Intl. J. for World Class Design to Manufacture*, March.

Sevenler, K., Sherman, M. K. and Vidal, R. (1993) Multidisciplinary teamwork in product design: some requirements for computer systems, *ICED*, pp. 343.

Sohlenius, G. (1992) Concurrent engineering, *Annals of CIRP*, **41**(2), 343.
Willcox, M. D. and Sheldon, D. (1993) How the design team in management terms will handle the DFX tools, *ICED*, p. 875.

# 3 Proposed Architecture for the New CAD Method

## 3.1 Introduction

The CAD method promotes the concurrent design of products, equipment and control systems. Although these three design activities cannot be completely simultaneous, Chapter 2 has shown that concurrent design allows for considerable overlapping between them.

The starting point of a design activity is defined by a set of hypotheses which can be modified later on by other activities: we call this set of hypotheses, and the pre-design which results from it, the *rough design*.

Let us give an example within the assembly sector:

- The *assembly planner* defines an assembly plan on the basis of the geometry of the product, whose accuracy relies on the information given by the designer! However, the determination of a better assembly plan could result in modifications to the designed parts.
- In the same manner, the *resource planner* produces a logical layout based on a rough estimate of the cycle time, which may later be modified by the *simulation module* when a more accurate measurement has been determined.

  However, the simulation module requires this rough estimate to start with a rough layout model.

This approach introduces a *dialectic* aspect in system design (Gonseth, 1926; Gonseth, 1947) that could be illustrated, speaking in general terms, by the two following steps:

- *Start from a set of hypotheses to establish a set of certainties.* In order to start as soon as possible instead of waiting for the final results of the preceding design step, each module must start with an incomplete set of hypotheses. As a consequence, the user needs to accept an initial phase during which *only a rough design* can be achieved. During this rough design phase, the user should anticipate the effects of potential changes due to the influence of other activities. Therefore, he or she can establish a series of *certainties*, which in turn may influence the hypotheses which are currently the basis of other activities.
- *Refine the design after the propagation of external influences.* This set may be enlarged and modified later on, thus requiring the design activity to adjust to these changes. In turn, these modifications may be propagated to other activities by updating the current set of hypotheses and certainties, and communicating the relevant ones to their consumer. During this phase, the user increases the precision of the design by increasing the number of *certainties*.

The success of the concurrent approach relies on:

- how large is the set of certainties compared to the initial set of hypotheses;
- how fast are the changes propagated;
- how fast a hypothesis is changed into a set of certainties.

Naturally, this requires data to be exchanged between the different activity modules. An efficient data exchange requires the relationship between the activities to be carefully defined.

This chapter presents a possible model of architecture for a new CAD method for industrial assembly, based on a model developed at the Ecole Polytechnique Fédérale de Lausanne (EPFL) under the name of the *Cubic Circuit* during previous research activities (Verdebout, 1993). Since this architecture presents many innovative aspects, it is presented in more conceptual terms and related to other solutions found in the literature (Scheer, 1988, etc.)

Generally speaking, the production of a product involves *three main flows*, which are further developed later on in this Chapter:

1. The *material flow*, of which the unfinished products and components are part of;

2. The *design flow*, which expresses the flow of concepts during the design of the products and equipment;
3. The *control flow*, which allows the running of the virtual models of a discrete events simulation, as well as the management of an actual production system.

These main flows can be represented as being the three orthogonal axes of a *production frame*, in which our cubic circuit is defined.

The expansion of the cubic circuit according to the design and control views forms the base on which the proposed architecture is actually built up.

It defines five *levels of reality* within which numerous secondary Cubic Circuits coexist. Each of them is significant for a particular design activity.

All the modules of the proposed CAD method, meaning all the software activities as well as all the data structures, can be represented within the five levels of reality.

This leads to an "ideal architecture" for the CAD activity in general. However, not all the modules of this ideal architecture were addressed during the collaboration of the authors of this book. Due to these considerations, the chosen architecture is slightly different.

Both the ideal and actual architecture will be presented, before concluding.

## 3.2 The Three Main Flows of a Productive System

### 3.2.1 The Material Flow

The objects produced by a workshop usually follow successive stages of modification of their characteristics or of their location, collectively called *actions*:

1. Each of them requires the objects to be made available through a *feeding*, which marks the beginning of the need to control their flow across the workshop.
2. Then, the objects are located at the resource(s) where the operation will take place: for this *transfer action*, numerous techniques are used, depending on many criteria.
3. The *operation* itself results in a modification of the characteristics of the object.

4. Making quality is of much greater concern than in previous years: this *inspection* involves a measurement of the obtained characteristics compared to a specification.
5. Finally, the inspected objects must be *delivered* as soon as possible; from this moment, the control is no longer under the responsibility of the workshop.

As a result, each operation can be decomposed into the five successive steps described above, each of them being an independent action.

The same steps are achieved if successive operations are necessary; however, since the objects will remain under the control of the same cell, the delivery after an operation and the feeding before the next one are merged with the transfer action.

Assembly is a little special, since more than one object must be merged: all these objects need to be located within the resource.

The quality after some operations can be so high that an inspection is superfluous: it has been "included in the design" of the triplet (object, resource, operation).

The preceding considerations can be summarised by Figure 3.1.

Any assembly process can be split into a number of "atoms" comprising the basic actions, connected according to the precedence constraints between the operations.

The usual structure of the resulting assembly graph looks like

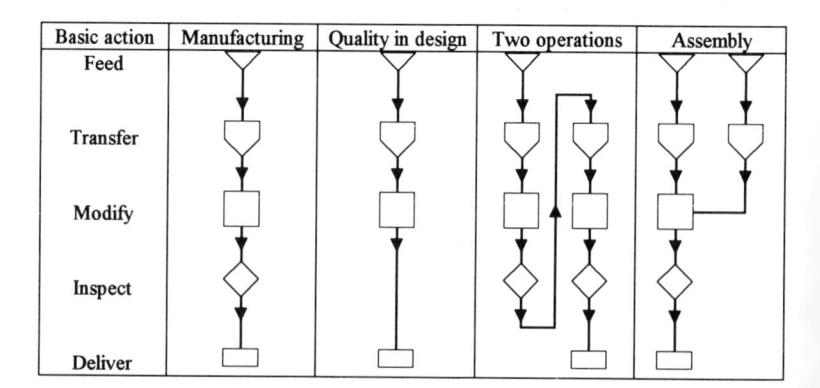

| Basic action | Manufacturing | Quality in design | Two operations | Assembly |
|---|---|---|---|---|
| Feed | | | | |
| Transfer | | | | |
| Modify | | | | |
| Inspect | | | | |
| Deliver | | | | |

**Figure 3.1** Several examples of representation of the material flow "atoms"

a tree, in which the root is the finished product, and the leaves are the objects comprising the bill of materials.

This concept of atoms, merged together if one considers them in an aggregated manner, has led the authors to consider a "fractal" model of the companies, as described in Warnecke (1992).

Naturally, this explicit representation is not always useful in everyday life:

- The feeding actions are represented in a particular manner (bills of materials).
- In many cases, the transfer actions are not explicitly represented.
- The sequence of operations is described in the manufacturing documents.
- The inspection actions are represented only when required in another document.
- The merging of a delivery and a feeding between two operations is not considered.
- When assembling one object (the component) onto another object (the receiver), the industrial plans define the component that results from a previous sequence.

In this simplified manner, the whole process can be represented by using only sequences of operations, thus indirectly considering the tree-like structure of the process.

We found it necessary to introduce this formalism for assembly systems, since:

- The feeding of components in automated assembly systems is one of the most expensive parts in the cost of the whole system.
- The transfer system represents an important part of the whole system's cost.
- The representation of the assembly operations by means of standard icons is an interesting tool for modern software (VDI 2860 has been used in our case).
- The management of the quality of products is an ever more important concept.
- In many cases, the delivery action also requires expensive machinery.

As a result, the resources of an automated or semi-automated assembly system may be devoted to any of the five basic actions, which leads to Figure 3.2.

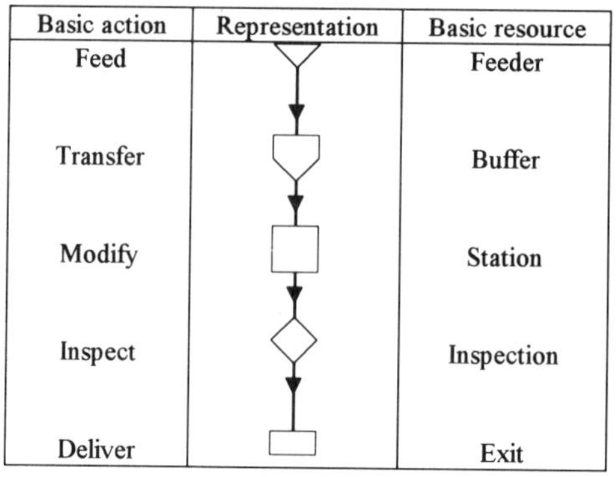

| Basic action | Representation | Basic resource |
|---|---|---|
| Feed | ▽ | Feeder |
| Transfer | ▽ | Buffer |
| Modify | □ | Station |
| Inspect | ◇ | Inspection |
| Deliver | ▭ | Exit |

**Figure 3.2** Correspondence between the basic actions and the basic resources

There are numerous categories of resources in everyday life, especially in assembly layouts, due to the fact that many different techniques and actions can be achieved. However, by representing such resources in an object-oriented manner, one can define an inheritance tree of the basic properties of such actions and resources.

Thus, the understanding of an assembly system and of the assembly process associated with it is dramatically simplified. In particular, this representation helps in the early transmission of the knowledge and hypotheses between the assembly and resource planning activities, on the one side, and simulation on the other.

One step later on in the design process, the same concepts will help a lot in the transmission of knowledge between the simulation activity, on the one side, and the control system on the other side.

This representation has helped in defining the behaviour of the assembly system for our concurrent design approach. Be it during resource planning, simulation or control of an actual system, even if the level of detail or accuracy of the data is different, some parameters will remain:

- representation of the same objects and resources;
- existence of the same actions within the process;
- precedence constraints between the operations;
- durations within the resources (cycle, set-up, maintenance, repair, etc.);

- performances of the resources (defects, jams, breakdowns, etc.);
- control of the flows of objects (batches, switching rules, etc.).

The main difference between the points of view of designers, simulation experts and layout control managers will be the "level of reality" of the objects they deal with. Each of these three categories of experts will consider the model of the layout as being a realised solution of another, or the representation of the problems of yet another.

*3.2.2 The Design Flow*

The design flow is naturally most carefully examined in our proposed CAD method, in which product design is achieved concurrently with the design of the resources.

The first design activity is triggered by the expression of a need for a given product, expressed in a company by the marketing activity, which will not be addressed.

Then, five successive steps will be achieved:

1. *Research*. The research activity provides technical solutions which can help in the design. This step leads to a set of materials and techniques possibly used in products. Its result is a "knowledge-based system", be it computer aided or not, which helps designers in choosing among the possible materials or techniques.

   On the other hand, this step decomposes the needed product into a set of functions to be achieved by the product to be designed, that is, the definition of the product. Design for Assembly is a part of this activity.

2. *Design*. The designers transform a definition into a set of characteristics of a product which satisfies it, including the points of view of the assembly department. Several points of interest are to be addressed:
   - computer aided design of the mechanical parts and of the resources;
   - assembly and resource planning.

   Most of these activities will be addressed within this book. This step results in a design of the product and resources, in the assembly plan to be achieved and a "logical" layout comprising all the necessary resources. However, this proposition relies on an estimate of the expected performances.

3. *Prototyping.* This step consists in improving the design by making prototypes of the resources and the product, to ensure the feasibility of the proposed solutions. The industrial department transforms both the assembly plan and the logical layout into a more precise work basis. Additional actions are planned in order to satisfy the technological constraints.

The assembly plan, based on the product design, is transformed into the assembly process, a graph of possible actions to be achieved by the chosen resources. This step also transforms the logical layout into a practical layout model. In the proposed CAD method, this step also validates the improved proposition by running a model which makes use of the simulation, flow control, scheduling and monitoring modules.

This helps in determining the proper decision making in these two last activities by testing the proposed solution against other possibilities. For example, the scheduling module will provide the user with a set of heuristics which can be evaluated, thus helping in making the proper choice, and aiding in the evaluation of other production constraints such as lot size, number of variants, etc.

4. *Controlling.* During the actual production, the user manages the system by using a supervisor on the basis of a layout model. The orders are transmitted to the resources by means of a communication system, which returns a representation of their current state.

The statistical production and quality control (SPC / SQC) data is computed in order to be presented in a more convenient manner, by means of time and quantity integrators. Supervision involves the flow control and scheduling modules which make use of the heuristics and methods determined during simulation.

At the resource level, the control module manages the production on the basis of the bills of work, and of the current resource state.

5. *Execution.* The last step in this hierarchical approach is the execution of the controller's instructions by the resources, as well as the measurement of any relevant data, to provide the controller with an accurate system state. This means that this step is achieved by the actuators and sensors of the resource.

These five successive steps are summarised in Figure 3.3.

# Proposed Architecture for the New CAD Method

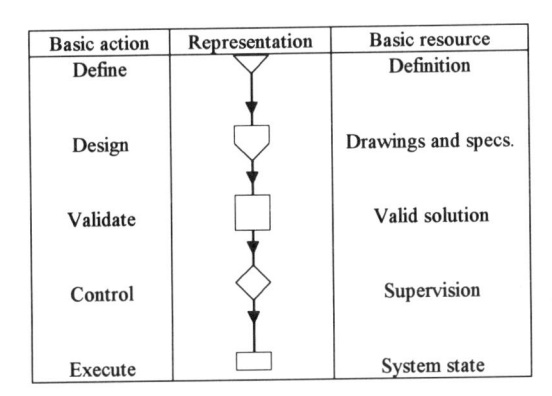

| Basic action | Representation | Basic resource |
|---|---|---|
| Define | ▽ | Definition |
| Design | ▽ | Drawings and specs. |
| Validate | □ | Valid solution |
| Control | ◇ | Supervision |
| Execute | ▭ | System state |

**Figure 3.3** The five basic actions in the design flow

In Figure 3.3, we have represented these five activities by the same symbols as in the material flow; by doing this, we express a homomorphism between these flows.

This is a first step towards a generalised model of the knowledge and mental activities which take place in an industrial company.

Such an epistemological approach by the construction of a model has helped us in defining the necessary blocks of the proposed architecture.

Naturally, this model is not always applicable: an actual system may be more complex and involve more intermediate steps.

## 3.2.3 The Control Flow

This section should be more familiar to the users of production planning systems.

This flow of information is usually separated into the following steps:

1. *Company planning.* Management defines the production objectives in the long term. This results in an investment plan which is assumed to satisfy the needs, but the details of the production to be achieved are not yet known.
2. *Production planning.* Production planning arranges the company objectives into a set of predictive production orders, based on the predicted distribution of the customer demands. This allows the assignment of operations to a proper set of resources.

This step also predicts the consumption of components by the workshop and pre-arranges the demand to other workshops of the same company ("make") or the purchase from other companies ("buy").

3. *Scheduling.* The scheduling module is responsible for sequencing the customer demands once they have been received by the workshop. It may split such demands, group others, arrange them in a proper sequence. Generally speaking, it converts a set of customer demands into another set of production orders.

    In turn, the orders are split according to the bills of materials in order to satisfy the needs for components (material requirement planning).

    It also manages the assignment of the jobs to the resources by exploding the structure of the manufacturing and/or assembly process into a series of bills of work to be sent to the resources.

4. *Order release.* The flow control module is in charge of releasing the production orders to the resources, assuming that the necessary components are present. It controls the execution in a global manner by monitoring the end of the execution of a given bill of work within a resource, in order to be able to release the next order as soon as possible. It also manages the flows of components and/or sub-assemblies within the workshop by means of a transfer system of any kind. To do this, it creates a set of switching rules that are to be obeyed by the transfer system during execution.

5. *Execution control.* The control system is in charge of triggering the resources themselves during execution, according to the bills of work they have received. This control module will trigger the cycle of the resources as soon as the previous cycle has been terminated in a satisfactory manner. This module will trigger the possible compensations due to quality problems if this procedure is to be achieved in real time: if a component is rejected, it must be replaced by a corrective action of the control module.

These five steps of the control flow can be represented as in Figure 3.4.

Once again, the same symbols can be used to represent this flow.

This is one more step towards a systematic representation of the whole activity of the workshop, be it from the point of view

# Proposed Architecture for the New CAD Method

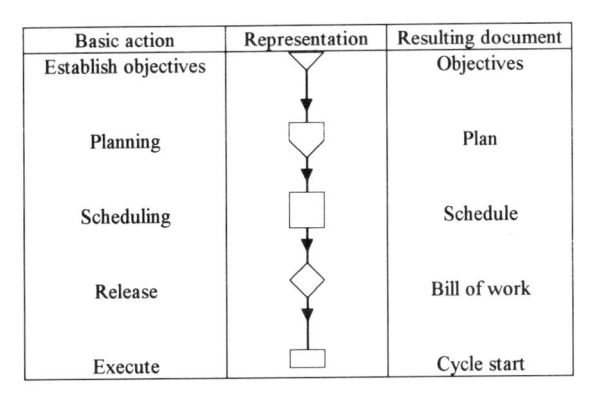

| Basic action | Representation | Resulting document |
|---|---|---|
| Establish objectives | ▽ | Objectives |
| Planning | ⬡ | Plan |
| Scheduling | ☐ | Schedule |
| Release | ◇ | Bill of work |
| Execute | ☐ | Cycle start |

**Figure 3.4** The five basic actions of the control flow

of the material, design or control flows. However, it must be noted that the time horizons are not the same:

- The material flow considers a very short time horizon.
- The design flow considers the long term.
- The control flow is of the intermediate term, when considering the two others.

## 3.3 The Levels of Reality in the Design Flow

### 3.3.1 The Production Frame

We saw in the previous section that three points of view coexist when one wants to represent the different activities of a workshop: the material, design and control flows.

However, these activities do not consider the same time horizon (or "term"). They can be considered as three orthogonal axes forming a production frame (Figure 3.5). This representation is closely related to the CIM architecture as defined by Scheer (1988), used in many references related to this topic (Mülkens, 1992), etc., as shown by Figure 3.6.

The comprehensive CIM model is presented in Figure 3.7. This CIM architecture expresses the manufacturing point of view, but it can be naturally transposed in our domain. It is, however, noticeably different, since for example the inventory control is placed between manufacturing and assembly, as if the

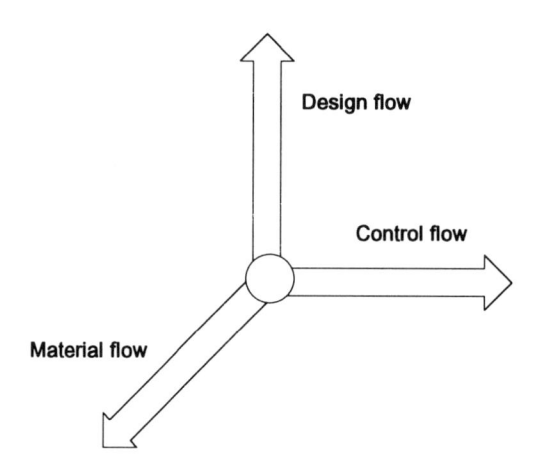

**Figure 3.5** The Production Frame representing the three main flows

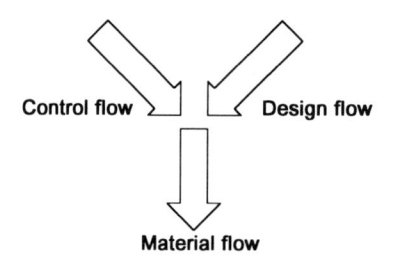

**Figure 3.6** Another view of the CIM architecture

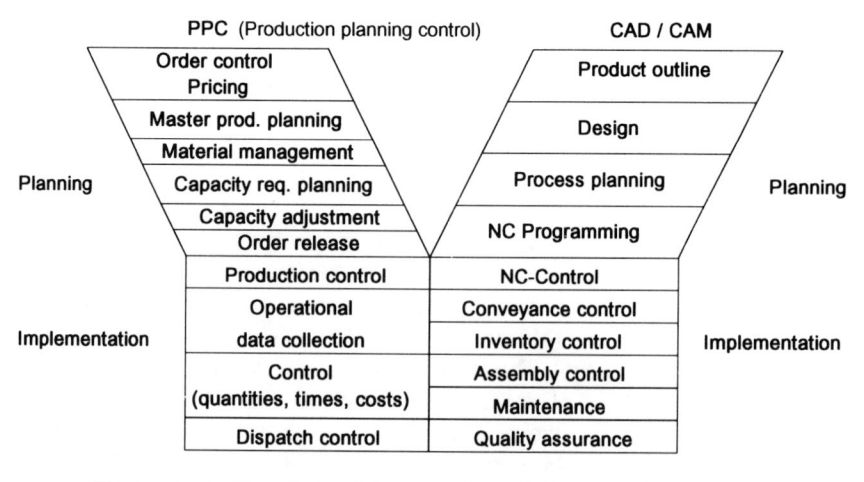

| PPC (Production planning control) | CAD / CAM | |
|---|---|---|
| Order control / Pricing | Product outline | Planning |
| Master prod. planning | Design | |
| Material management | | |
| Capacity req. planning | Process planning | |
| Capacity adjustment | NC Programming | |
| Order release | | |
| Production control | NC-Control | Implementation |
| Operational data collection | Conveyance control | |
| | Inventory control | |
| Control (quantities, times, costs) | Assembly control | |
| | Maintenance | |
| Dispatch control | Quality assurance | |

**Figure 3.7** The CIM architecture (from (Scheer, 1988))

# Proposed Architecture for the New CAD Method

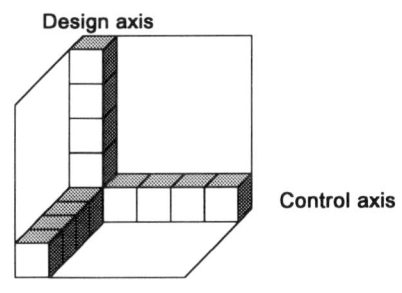

**Figure 3.8** An activity considered from the point of view of the three axes

manufactured parts were always to stay in a warehouse for a while, which is not always the case.

The origin of the production frame in our model is also very interesting and will be developed in the next section.

Each activity of the production system can be considered and developed from the point of view of the three axes, as shown in Figure 3.8. As a result, each activity can be split into $5 \times 5 \times 5 = 125$ blocks! Naturally, some are much more important than others.

The most important thing here is to see that this model allows for the splitting of an activity into five successive or concurrent steps, at least according to the control and design flows. From the design point of view, the activity is split according to five successive achievements in terms of design. The activity is more and more concrete, meaning that it achieves five successive steps in its realisation (definition, design, industrialisation, validation, production).

From the control point of view, the activity is organised in five successive steps of aggregation:

1. The production of a company is arranged as a set of product families.
2. The product family is organised as a set of variants.
3. A variant is an assembly; therefore, it is organised as a set of assembled parts.
4. A part is a set of shapes or any kind of characteristics, including the material it is composed of.
5. The shape or characteristic is the elementary primitive of product design.

### 3.3.2 The Cubic Circuit

This section intends to define the elementary activity of an industrial system.

To define this model, since we want to represent the activity of a deterministic system, we use the epistemological approach as presented by the Swiss philosopher Jean Piaget (Dolle, 1991). Two concepts will be used:

1. the model of an activity, of a mental or computing process, represented by a cube;
2. the model of the result of an activity, or of a perturbation of the system, or of a data structure or any other kind of knowledge, represented by a sphere.

We have searched for a minimal representation of a deterministic system, that is, of a system which tends to "survive" by reacting in an adapted manner to the perturbations. Let's imagine that such a system receives a perturbation from the external world.

A customer demand can be considered as being a (welcomed) perturbation coming from the external world into the assembly system. Being deterministic, this system tends to satisfy this demand, or in more general terms, to compensate or balance this perturbation. In order to achieve this, the system will successively:

1. react to this perturbation by emitting a production order, that is, process a command, which translates the intention of satisfying the demand;
2. react to this order by expressing a need, which in turn will interact with the external world through the perturbation of another system (for example, purchase components);
3. once the need is satisfied, react by using its own resources to provide a change in the characteristics of the components—this is the operation or production itself;
4. once the characteristics are modified, check by an inspection that the resulting changes actually satisfy the expressed demand.

This summarises the actual activity of an industrial system:

- At the reception of a customer demand, the system controllers will emit an order.
- The system will then express the need for a set of components and make them available.

- Once the components are made available, the system resources will modify them to assemble a product.
- Then, the system will inspect the characteristics of the resulting product in order to check their conformity with the customer demand.

To ensure its own survival, the industrial system needs to generate a profit: this will ensure that the benefit of a sale is higher than the cost of the purchased materials and operations. That is, the system needs to ensure that the lost resources are lower than the gained resources, so that the wealth of the system ever increases (sorry about these drastic simplifications!). This elementary model can be represented by a graphical model that we call the Basic Cubic Circuit, as shown by Figure 3.9.

A central processing block, not represented in Figure 3.9, would represent the mental activity of *coordinating* the other activities. It is translated in Figure 3.10 by its triggering events and their results.

The advice-based processing blocks (design for assembly, assembly and resource planning) will emit advices to the user, who in turn will interact with the modules. The simulated world, as well as the supervisor model, will interact with the user by means of events, simulation results, etc.

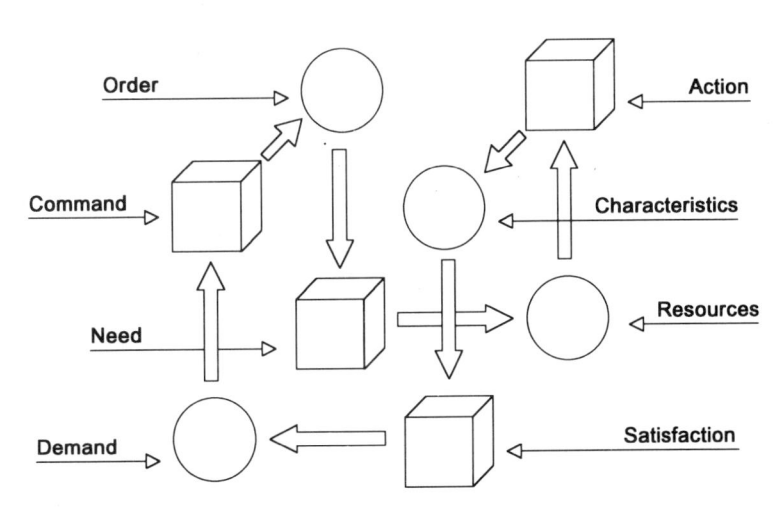

**Figure 3.9** The basic Cubic Circuit

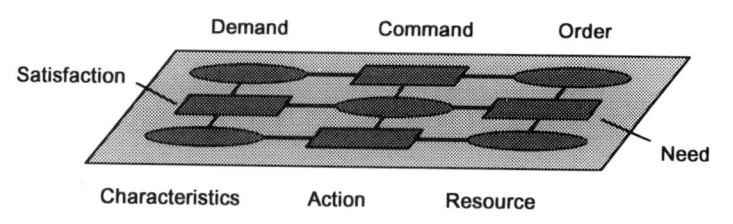

**Figure 3.10** A plane view of the same cube

### 3.3.3 The Five Levels of Reality

After splitting the elementary activity or data structure according to the three points of view (material, design, control), and determined the basic cubic circuit, it seems interesting to merge these two models into a synthetic model of system activity.

In Figure 3.11, we show how this model can be established; for the sake of clarity, only the design and control views have been developed. The same shades of the rounded rectangles of the data structures, as well as of the rectangles of the activities, indicate an identical level from the design point of view, that is, an identical level of reality within the model.

It is possible to represent all the modules of the CAD method for industrial assembly in a matrix according to the model of Figure 3.11, but the resulting figure is not easy to interprete in terms of basic cubic circuits.

Instead, we preferred to split the layers by using the plane representation of the cubic loops according to the five levels of reality as displayed in Figure 3.12.

Each level of reality is explained in the next section.

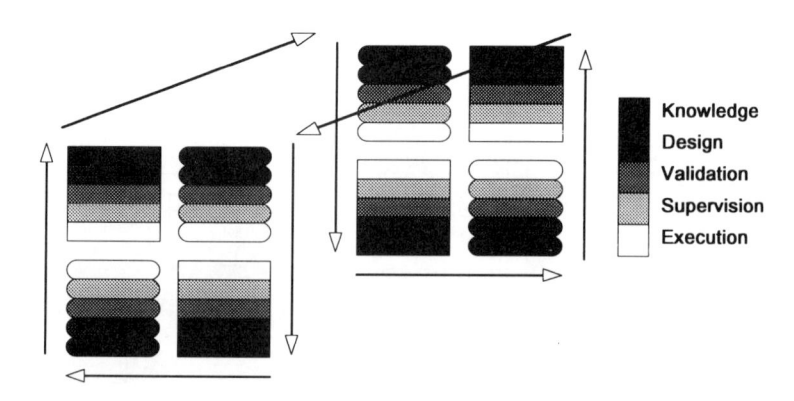

**Figure 3.11** Expanded Cubic Circuit model

# Proposed Architecture for the New CAD Method

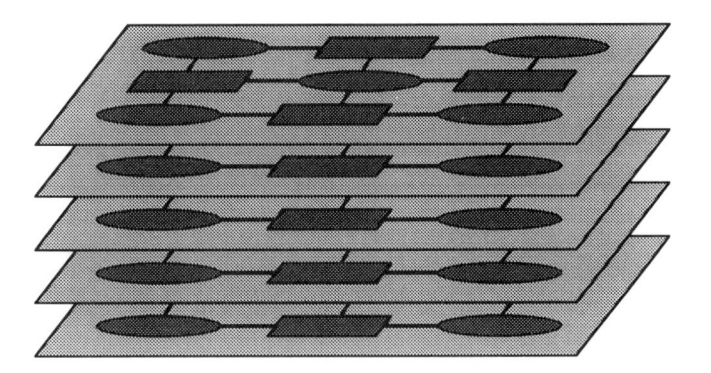

**Figure 3.12** The cubic circuits in the five levels of reality

In our opinion, the most important result of this model is the considerable number of relations between its different blocks.

In the usual CIM architecture as described in Scheer (1988) and by many other authors, the three flows were described as being successive activities. This naturally induces relationships between the successive planes of our model, that is, between the levels of reality.

The connections between the blocks of a same level are usually described by means of a very large and central data base, a kind of "big brother" through which people are expected to communicate. The modules are connected to their neighbours only by means of successive abstractions or instantiations, which are represented in our model by the vertical connections. Furthermore, our model adds extra relationships within the plane itself, thanks to the concurrent approach of the design problem.

Each plane is an *abstraction* or an *instantiation* of its neighbours; in addition, it reproduces all the productive functions of the company at a given level of reality.

Another advantage of this model is that it expresses four databases in each level of reality, each of them being another view of the same problem (see Figure 3.13).

Finally, the addition of the deterministic aspect is a very important topic since it brings the human creativity back to the centre of each level of reality of the model, as opposed to traditional CIM approaches.

In the same manner as the survival instinct separates living

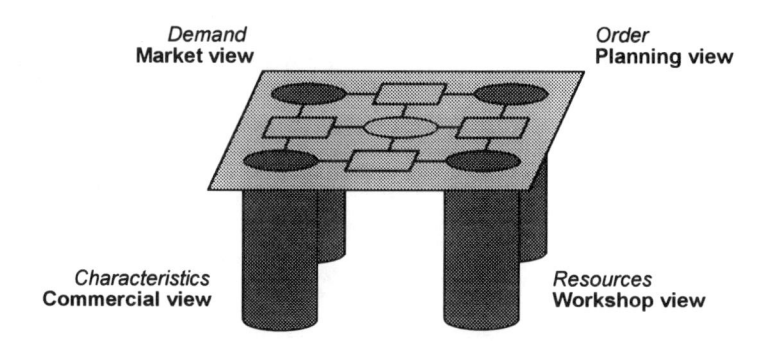

**Figure 3.13** Data bases and points of view in the cubic circuit

- Demand / Market view
- Order / Planning view
- Characteristics / Commercial view
- Resources / Workshop view

creatures from inanimate objects, the human coordination in its five aspects is the expression of the company's creativity.

The CAD method proposed in this book is not an automated design tool but a computer-aided design tool for the user; the advice-based modules being a perfect expression of this need.

## 3.4 The Ideal and Chosen Architecture for a CAD Method

For each level of reality as presented in the Section 3.3.3, a cubic circuit can be established. Since the model is consistent and complete, we can derive an ideal architecture of modules for a comprehensive computer-aided centre of activities. Such an architecture would gather the point of view of all the people involved in production:

- product and resources design (CAD);
- process planning and scheduling (CAPP);
- manufacturing and assembly (CAM).

The chosen architecture is essentially the same, except that not all the modules were treated during the project. The modules that were not treated are not heavily shaded in, in Figures 3.14–3.18. Furthermore, the scope of the project can easily be visualised or represented in the five levels of reality of the model.

### 3.4.1 The Knowledge Level

This level of reality is the most conceptual one. It will only be partially dealt with in this book:

Proposed Architecture for the New CAD Method

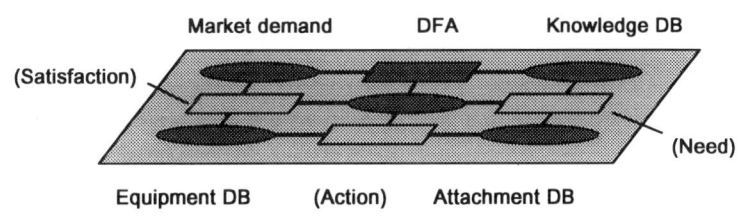

**Figure 3.14** The cubic circuit at the knowledge level

- Only one processing block will be described (design for assembly).
- Market demand is supposed to be known by means of a product concept to be provided by the marketing department.
- The knowledge database expresses the design rules which will be followed during the design of the product according to the specifications.
- The attachment database expresses the technological knowledge of the company in terms of assembly.
- Finally, the equipment database expresses a catalogue of the available resources within the company, implying that the design of a new resource is not necessarily foreseen.

The expression of the determinism at this level is the emission of advice from the DFA module to the user.

## 3.4.2 The Design Level

This level (Figure 3.15) is responsible for proposing a possible solution to the product specifications, according to:

- the knowledge which was accumulated in the plane above, especially from DFA;
- the knowledge accumulated at the simulation level.

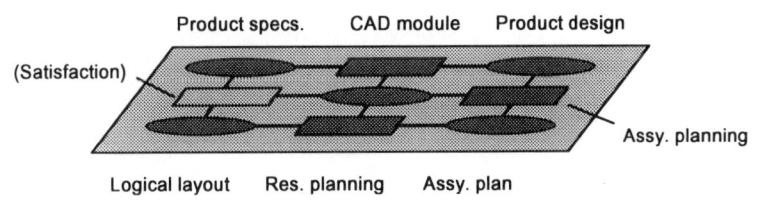

**Figure 3.15** The cubic circuit at the design level

Our CAD method for industrial assembly presents a consistent and powerful architecture at this level:

- The specifications are transformed into a product design by means of existing CAD tools connected to other modules.
- The innovative assembly and resource planners help the user in defining the assembly plan of the product and a logical layout which comprises all the necessary resources.

Once again, two modules of this level interact with the user by emitting advices, in addition to the usual interactions with the user, as in a modern CAD environment.

### 3.4.3 The Simulation Level

This level (Figure 3.16) interacts with the level above by accepting its rough design as a starting point for the simulation models, thus allowing an early start of the activity in a concurrent approach.

It also interacts with the supervision level by updating the layout models, as well as the estimates of all the parameters of the resources and operations.

Its other originality, compared to other commercial products based on a simulator, is the connection of innovative flow control and scheduling modules.

Both of these modules will interact with the user exactly in the same manner, be it off-line or on-line, thus helping designers in acquiring a better understanding of the workshop needs.

One remark on the model: the arrangement of the modules helped the SCOPES partners in understanding that the flow control module expresses the needs of the workshop for updated and scheduled orders.

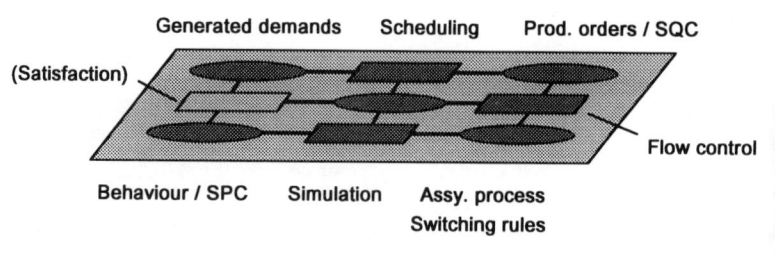

**Figure 3.16** The cubic circuit at the simulation level

In the usual CIM approach (Scheer, 1988), this is the purpose of the order release module.

The satisfaction module can be easily added to the current architecture, in order to comply with the need for an optimal layout.

The simulation module described in this book could easily be completed with a scripting facility which would help the user in defining a powerful optimisation tool.

The determinism at this level is achieved by means of events, which trigger predefined reactions of the other modules.

The interaction with the user consists in defining these reactions: during simulation, the user will validate them by observing the simulation run, interacting with the simulator, and analysing the simulation results.

### 3.4.4 The Control Level

This layer (see Figure 3.17) is connected to the simulation level and takes advantage of simulation results by downloading the finalised models and chosen strategies. Naturally, these models may be improved upon with day-to-day experience that is acquired during production.

It is also connected to the workshop level by means of the communication hardware and software; once again, the authors of this book took great advantage of existing experience in defining the features of a powerful supervisor. However, by using the same innovative concepts of flow control and scheduling as at the simulation level, the interaction with the user is improved; the similarity with the architecture of the level above is remarkable.

Many specifically expressed users' needs could be fulfilled by applying the proposed solutions, which improve workshop flexibility in industrial conditions.

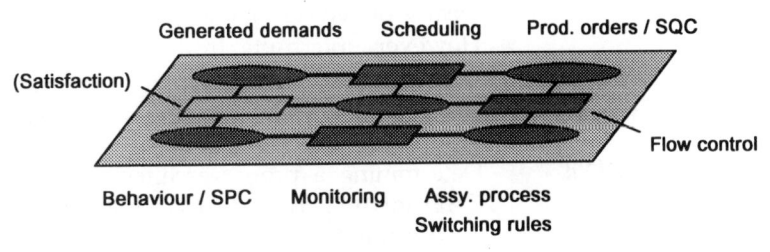

**Figure 3.17** The cubic circuit at the control level

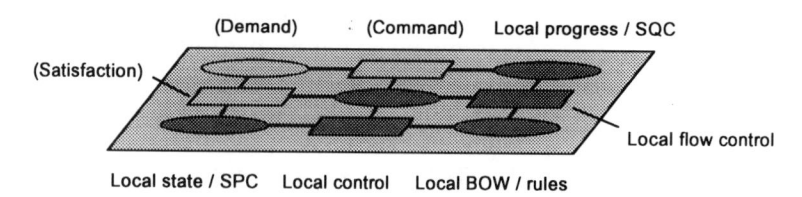

**Figure 3.18** The cubic circuit of the execution level

The determinism at this level is similar to that of the simulation level, except that events are exchanged between different controllers, or external modules, such as a backup scheduler.

Events are also massively used between the supervisor and the controllers which manage the execution level.

### 3.4.5 The Execution Level

This level (Figure 3.18) is connected to the control level by means of the communication module. The innovation at this level consists in decentralising the intelligence whenever possible, thanks to the increasing computational capabilities of today's controllers. Such a decentralisation can be achieved by means of standard software parametered by the supervisor.

As a result, the necessary investment in dedicated software is dramatically reduced. System safety is also improved by providing an autonomous mode to the resources if the supervisor is down. Tools are also provided that ensure safe human intervention within the resources when necessary.

At this level, the determinism is also achieved through user interaction and by means of events in the local controllers, and their associated Petri net-like programming.

The exchange of data between this level and the supervision level is achieved through the communication network. The standard method is for the supervisor to control this level by means of asynchronous polling, which limits network loads. However, interrupts from the local controllers may be used when required.

### 3.5 Conclusions

Determining a proper architecture of the modules within our CAD method for industrial assembly proved to be one of the most difficult tasks.

Each of the partners initially had a strong point of view due to his own experience in one or several specialised fields, thus perfectly illustrating the difficulties in a concurrent engineering approach!

The purpose of this section was to present a possible explanation of the chosen architecture by means of a model, which is naturally a simplified, but hopefully satisfactory, representation of the industrial realities.

Lots of literature has been written since Scheer (1988) to avoid the drawbacks of this pioneering work: let's mention an interesting survey with many references in Williams (1994).

The current representations of the CIM architecture proved to be insufficient to represent our concurrent approach. We hope that the proposed architecture will also contribute in clarifying the general understanding of this field. We applied a more epistemological approach with the help of the concept of determinism in the industrial activity. Research in this field is still wide open.

The sections that follow present all the modules of the integrated CAD method in more detail.

## 3.6 References

Dolle, J. (1991) *Pour comprendre Jean Piaget*, nouvelle édition mise à jour, Privat, Toulouse.

Gonseth (1926) *Les fondements des mathématiques*, Blanchard, Paris.

Gonseth (1947) L'idée de la dialectique aux entretiens de Zürich, *Dialectica* no. 1, p. 35.

Mülkens (1992) *Cours de Gestion de la Production*, LGP-EPFL, Lausanne.

Scheer (1988) *CIM Computer Integrated Manufacturing, Towards the Factory of the Future*, Springer-Verlag, Berlin.

VDI 2860 *Assembly and handling; handling functions, handling units, Terminology, definitions and symbols.*

Verdebout (1993) *Vers des automatismes de sécurité pour l'assemblage automatisé*, lecture given at Télémécanique Nanterre, internal publication, EPFL, Lausanne.

Warnecke, H.-J. (1992) Unter Mitw. von Manfred Hueser—*Die Fraktale Fabrik, Revolution der Unternehmenskultur*, Springer.

Williams, T.J. *et al.* (1994) Architectures for integrating manufacturing activities and enterprises, *Computer in Industry*, Vol 24, Elsevier, Amsterdam.

# 4 Product Design for Assembly

## 4.1 Introduction

Companies today operate in increasingly competitive environments. As well as competing on the price of their product, they are also competing on the quality and the speed with which they can introduce a new product to the market. It has been estimated that as much as 40–60% of the total product cost is accounted for within the assembly of complex products. Designing products that are cost effective to assemble gives an important competitive advantage. Techniques and methodologies to analyse products for ease of assembly are increasingly being adopted.

Design for assembly is a key part of concurrent engineering. Concurrent engineering is often seen as gathering together designers, manufacturing engineers, process monitors, marketing personnel to work in teams at the pre-design stage. The teams need to be supported with quantified cost data and a systematic design evaluation process. A product analysis for assembly early in the design process provides the teams with a systematic approach to evaluate design concepts.

The purpose of design for assembly is to reduce the product development time and cost, and at the same time improve the quality of the product. In most typical mechanical and electromechanical products, the five main foci for assembly analysis are: simplification of the product by eliminating redundant parts within the design, and identification of part characteristics that would cause difficulties in handling, feeding, positioning/insertion or joining. A substantial body of design and assembly research can be traced to the original work in the early eighties by Professors Boothroyd, Swift, and Redford currently of the Universities of Rhode Island, Hull, and Salford.

With the rapid development of manufacturing technology and tools, designers can sometimes be unaware of alternative manufacturing and assembly processes that are beyond their normal practice and experience. Most successful case examples in Design for Assembly use processes that simplify the design through integration of parts. Metal casting, plastic injection moulding, sheetmetal bonding and forming, high speed machining, and superplastic forming/diffusion bonding are typical techniques to replace fabrication and fastening. By discussing the design with a manufacturing engineer, methods by which the design can be altered to improve its efficiency in terms of ease of assembly and manufacture can usually be found.

The product design for assembly (DFA) module in the integrated CAD method is an essential module in the concurrent engineering scenario. It provides the designer with redesign advice in order to facilitate easier assembly of the product. The module has links, directly or indirectly, to all the other off-line modules and the simulation module. DFA analysis is knowledge intensive. Appropriate alternatives for redesign are very specific to the particular industry and company. This chapter presents the specification and development methodology of a DFA module that satisfy the requirements of a group of European

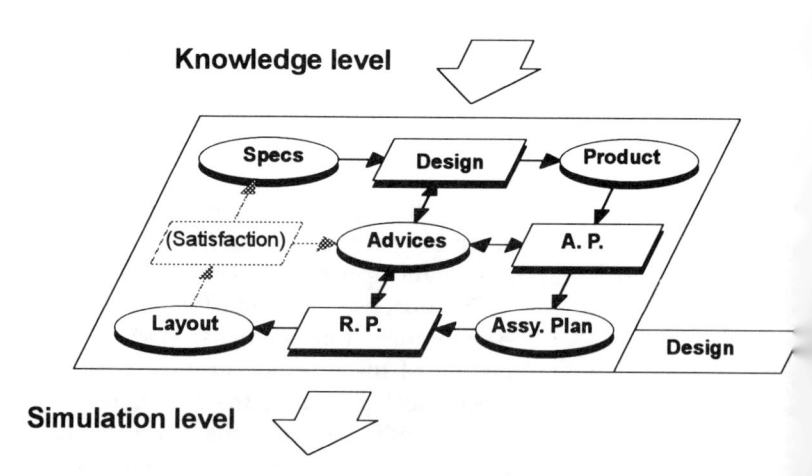

**Figure 4.1** Situation of the module within the CAD method for industrial assembly architecture

industrial manufacturers in the small electromechanical and large mechanical product domains. It can be used as a template for companies to develop their own DFA systems.

Product design is at the Design Level of the architecture. The realisation of the product is conducted through the intense exchange of information and knowledge on the capability of the assembly processes and the physical resources. They interact to result in the satisfaction of the product specifications. Supporting the Design level is the Knowledge level that stores the knowledge for Design for Assembly, assembly actions and resources (see Section 3.4.1). The output of the design level feeds the Simulation level of the architecture.

The Product Design For Assembly module aims to provide positive advice to the user on how a design might be improved to ease its assembly throughout the design process. This support is context sensitive and is available through the concept design to detail design stages. The DFA module consists of the five submodules:

1. product structure
2. handling
3. feeding
4. positioning and insertion
5. joining.

These can be accessed during any stage of the design process.

Each module requires a different set of data inputs and a different set of rules. The data input requirements, the analysis order and the set of applicable rules for each module are explained in the appropriate sections.

## 4.2 Users' Needs

A detail study of the application and requirements of design for assembly in European industries generated a list of characteristics for DFA support (SCOPES D3, 1992). The study was based on electromechanical contactors, assembled on a automated high volume assembly line; and machine tools which are low volume, large complex products, assembled manually in a job shop environment.

The general conclusions on DFA requirements are:

- At present systematic DFA is infrequently used during the design process. Design for manufacture and assembly is con-

ducted through liaison between the design and manufacturing departments. DFA tools must be easy to use and widely available, in different forms, throughout the design process. The most cost effective use is at the beginning of the product/process development.
- DFA as a design support tool must be fully integrated with the assembly planning, resource planning and simulation modules. These modules share common product and process data that should not be duplicated.
- DFA tools must be customisable to the users requirements to take into account knowledge that is company and product specific. Company specific capability and constraints are needed in addition to general advice and guidelines.
- The user should be able to trace the reasoning that leads to the DFA advice.

These provides guidelines to the design of an industrial friendly DFA tool.

The product specific characteristics for the study are:

- Average number of subassemblies ranges from four to forty.
- Average number of components in each subassembly, including the fastening means, ranges from 14 to 250.
- Number of product variations ranges from 8 to 100.
- The majority of assembly operations are a form of insertion. A check should take place to ensure that this insertion is made as easy as possible.
- The joining analysis should be capable of dealing with the following fastening methods: screws, nuts, bolts, rivets, fastening clamps, locking flaps, cable ties, adhesives, welding, wrapping, pressure fits and shrinking.
- A handling and feedability analysis is needed to reduce the problems associated with automated assembly.
- The most common part attribute which affects its assembly is the cleanliness of the component. Others, which are less important, are its cost and its flexibility.
- When dealing with the functionality of components, the following are the most common attributes: shaft, bearing, gear, valve, seal, fastener, casing, pipe, electrical and electronic.
- When trying to deal with geometric information, the most common features are: boss, pocket, through slot, open step, through hole, face, taper, thread, knurl, chamfer and fillet.

# Product Design for Assembly

Intelligent analysis rules are developed to support the design check of these product specific requirements.

## 4.3 State of the Art

A review of current systems and work on design for assembly is presented in the five areas of product structure, handling, feeding, positioning/insertion and joining. The current market leader in design for assembly system is Boothroyd and Dewhurst Inc.

### 4.3.1 Product Structure

Boothroyd and Dewhurst (1987) and Lucas (Swift and Doyle, 1987, Leaney and Wittenberg (1992) concentrate on highlighting parts that are not fundamental to the function of the product and trying to eradicate these parts. The Hitachi method (Bedworth et al., 1991) concentrates mainly on the assembly operations and does not check the product structure. Holbrook (1988) developed a system that checked whether standard components were being used. Shi (1991) implemented the DFA analysis in a CAD system and questions the user two parts need to be separable for maintenance or replacement, made of different materials, move relative to each other and whether they can be assembled in any order. Should the answer to all these questions be yes, then the system suggests that the two parts be united. The facility exists for the automatic combination of the two parts.

### 4.3.2 Parts Handling

The Hitachi method takes no consideration for handling or feeding parts. Boothroyd performs an analysis for the manual and automatic handling of parts, taking consideration of the weight, size and geometry of the part. Lucas assesses a part for its suitability for automatic feeding, gripping. Kim and Bekey (1990) developed a system that identified handling problems by counting the number of surface pairs of a part that could be grasped, the number of flat/non-flat surfaces of a part and the friction of these surfaces.

### 4.3.3 Parts Feeding

The Hitachi method provides no direct analysis for feeding, but both Boothroyd and Lucas consider a parts size, and its likelihood of shingling, tangling or nesting. Kim et al. (1992) have developed an axiomatic method to analyse the feedability of a part. At present this method is only suitable for the analysis of automated grasping and orientating by checking such factors as the number of graspable surface pairs of a part and the friction of each surface. Atiyeh (1992) has highlighted the fact that reducing the number of components and combining the functionality of components it becomes too difficult to feed them and assembly becomes more difficult rather than easier.

### 4.3.4 Insertion Analysis

The Hitachi, Boothroyd and Lucas methods all perform an insertion analysis by considering the direction of insertion and the difficulty in performing the insertion caused by the tolerancing and the chamfer present. Shi checks for all "peg-in-hole" insertions and checks whether a chamfer is present and the tolerances of the hole and peg. The accuracy and repeatability of the robot performing the insertion is considered. Should the insertion prove too difficult then an automatic redesign of the chamfer and tolerances is performed. Should the redesign prove unsuitable then it is suggested that the assembly robot is unsuitable.

### 4.3.5 Joining Analysis

Hitachi method analyses the fastening method indirectly, by looking at the assembly operations and relating their difficulty to a simple downward motion. Boothroyd and Lucas highlight the fastening parts as non-functional and hence eligible for redesign. The designer is therefore encouraged to reduce their number or select an alternative joining method. Nieminen *et al.*, (1989) have developed a feature-based system that interacts with the designer to determine the parameters for the automatic generation of standard screw and nut type joints by determining the joining feature between two components.

### 4.3.6 Integration of DFA with CAD Systems

Holbrook suggested that as the established DFA methods require that the designer answers a multitude of questions on each component within the assembly, the way to reduce the time for the DFA analysis would be to extract as much information as possible from a CAD model of the product. The solid model probably best describes the product from the geometry and attributes of individual components and their relationships.

Shi used this approach and developed two software modules to perform a design for insertion analysis and a parts redundancy analysis.

The Lucas method has been developed so that it can now, in prototype form, extract information from 2D CAD data and apply it through a knowledge-based system incorporating manufacturing, feeding and handling information. Li and Hwang (1992) have partially developed a framework for an automatic DFA evaluation system. They have partially implemented this framework with a feature-based system by developing an assembly features extraction algorithm to convert the CAD data into assembly features.

The product Pro/DFA was launched in 1994 as an implementation of the Boothroyd and Dewhurst methodology in the Pro/Engineer CAD system.

## 4.4 Functionalities and Methodology

### 4.4.1 Introduction

The product design for assembly module includes a range of analysis to support the product design and development cycle at different stages. The five main analyses are: product structure, handling, feeding, positioning and joining. The detail of the analysis is described in the relevant sections.

Information about the product and the manufacturing processes are incrementally defined in the design process. It is important that the concurrent engineering teams can receive support at the appropriate stages of design. The functionalities of analysis is tailored to the maturity of design. The complexity and richness of analysis increases as more information is available. The stages of application are:

*Stage 1.* At this starting level when a new component is to be designed, a check is made to see whether a component that already exists does not fulfill the same function. The geometry of the component is not required. It is possible to proceed with only a bill of materials (BOM). If a modular component is to be designed then an existing similar component is to be displayed with the relevant handling, location, etc. features highlighted.

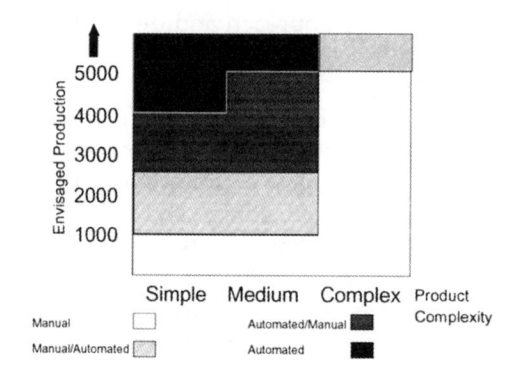

**Figure 4.2** Process selection chart

*Stage 2.* Once a new component has been designed it can be checked for its suitability for handling or feeding. The likelihood of whether it will be automated or manually handled and fed can be determined from a method determination graph or by user preference. The method determination graph (see Figure 4.2) is customisable by the user so that it may be tuned to a specific company.

*Stage 3.* Once two or more components have been designed and placed together an assembly results. Consideration

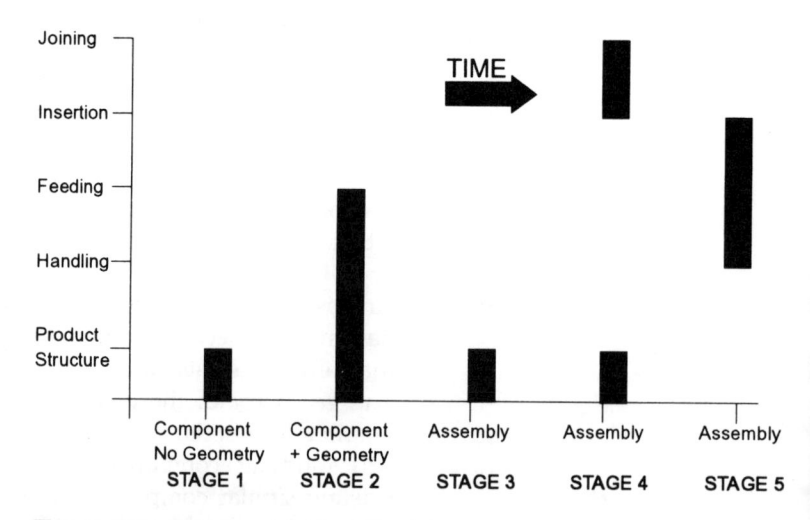

**Figure 4.3** Design support applications

can be given to modular design, the use of standard sub-assemblies and part reduction.

*Stage 4.* The insertion and joining operations in the assembly can be checked.

*Stage 5.* The handling and feeding of the assembly can be checked, as in Stage 2.

At each design stage and as the analysis progresses different combinations of data, with increasing levels of detail are required (see Figure 4.3). The users have to provide these data from knowledge, estimates or other sources. In an integrated CAD system, most of the data could be obtained from the geometric model of the product or other analysis modules in the system. The more data that is available from the CAD system and other integrated modules the less user involvement is required.

The following sections describe in detail the purpose of each module, the analysis order scenarios, the input requirements and where each piece of information can be obtained.

### 4.4.2 Product Structure Analysis

By considering the product structure, advice can be given on how the design can be rationalised across product families and modules by using the same components and subassemblies. This gives economy of scale for component production, simplifies process control and reduces equipment and tooling costs.

Proper configuration of the product into sub-assemblies allows division of design tasks and teams, and the accumulation of specific expertise.

Modularising the design and arranging for product variants to be produced as late in the assembly sequence as possible enables controlled variation and aids just-in-time production.

Standardising components and materials leads to fewer complications in purchasing, inventory management, tooling and manufacture. By purchasing off-the-shelf items rather than manufacturing them, the benefits of mass production even for products manufactured in low volumes are obtained.

To aid the selection of standard components, an easy way to browse through standard part and component libraries would prove very useful.

The product structure module is split into three sub-modules. The first determines the functionality of a component and which surfaces of a component are functional and vital to part performance. This is part of deciding whether a part is theoretically essential or not and hence whether this part can be removed from the redesigned assembly. The second aims at reducing the parts within the assembly by suggesting that they are combined if specific criteria are met. The final module checks components to see whether standard parts are available that would fulfill the same purpose.

**(i) Analysis order**

1. The functions of the parts are determined ⇒ This could be obtained from the class and feature reference library. If the reference could not be established, the user has to provide the functional description of the part (see Section (ii)).
2. From the functionality determined at step 1 suggest suitable standard parts ⇒ The user can either accept the standard part and place it within an assembly or design a new part (see Section (iv)).
3. Once two parts exist then the part redundancy criteria can be applied (see Section (iii))
   (a) Do parts have to move relative to each other?
   (b) Do parts have to be made of different materials?
   (c) Do parts have to be separable for maintenance or replacement?
   (d) Would the joining of two parts prevent the assembly of another part?
   ⇒ Should all the criteria be met then the suggestion would be made that the two components are combined.

1 = Biggest part
4 = Smallest part

Analysis Order
1 - 3
1 - 4
2 - 4

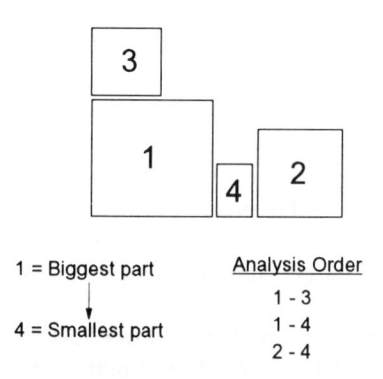

**Figure 4.4** Order of part analysis

# Product Design for Assembly

When checking an assembly, the assembly tree (see Section 5.4.4.(i)) should be used, if available, to guide the analysis order. This is also true for the insertion, location and joining analyses. Another sequence is the size of the parts. The largest part is analysed first followed by the next smaller part, and so on (see Figure 4.4). The functionality of the components is also used when determining the analysis order. Fasteners are analysed last as the aim is to reduce their number. If the base component can be identified, either by the user or from the CAD system, then the analysis begins with this part. Other suggestions for selecting the order of analysis are mass and volume envelope of components.

If two parts are combined and the parts redundancy procedure has been selected on its own then the part handling or part feeding module should be invoked next to check that if the two parts are combined it is still possible to handle or feed them.

**(ii) Functionality analysis**

**Rules for functionality analysis**

The prioritised functionality rules are listed below.

1. If an in-house designed component performs the same function as a standard component then consider using the standard component.
2. If a tolerance is attached to a surface then make the tolerance as large as possible without affecting functionality.
3. If a tolerance is high then the surface is probably functional
4. If a surface performs one of the following:
   — Is the base component of an assembly
   — Is a joining interface
   — Transmits motion
   — Transmits forces
   — Aids handling
   then the surface is functional and should be labeled as such.

**Functionality data requirements** The data inputs are required by the module and have been determined from the above set of rules. An explanation of how each input is used is provided. The source that is likely to provide the input is also listed.

| Input | Provided by |
|---|---|
| Component functionality | User |
| Base component of assembly | User, assembly planner, CAD |
| Base area and centre of gravity | CAD |
| Tolerances | CAD |
| Surfaces that transmit force | CAD |
| Surfaces that have motion | CAD |
| Surfaces that provide location | CAD |
| Features that aid handling | CAD |
| Fastening features | CAD |

- *Component functionality.* A new component's functionality should be identified by the user before the component is drawn. This will enable existing standard components with a similar functionality to be identified to the designer. This could save the design of a new part. The most common functionalities are:

shaft, bearing, gear, valve, seal, fastener, spring, pin, casing, pipe, wire.

These should be available from a tree like list from which the designer can refine the selection, and pinpoint the exact functionality required. The company should also be able to update the list to tailor it to its product range.

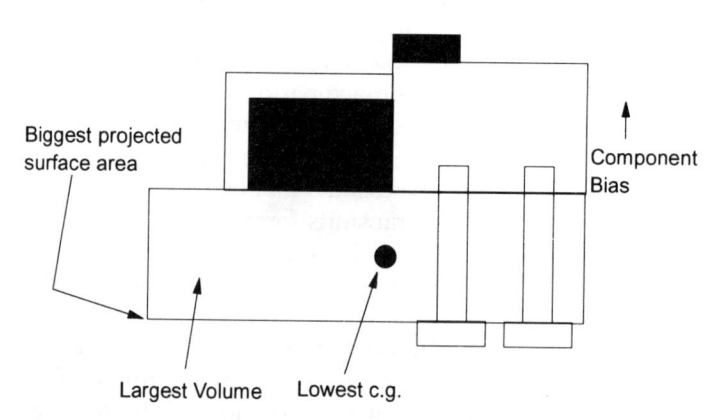

**Figure 4.5** Determination of base component

- *Base component of assembly.* The base component of the assembly can be identified from the assembly planner (see Section 5.4.2.(i)), by the user or directly from CAD. The volume, the centre of gravity (c.g.) position, the projected surface area and the component bias may all be used as suggested as possible base components (see Figure 4.5). Once the base component has been identified then this may used to determine the analysis order for an assembly.
- *Base area and centre of gravity (c.g.).* Once the base component has been identified, the area of the base can be determined and the vertical position of the c.g. located. The stability of the base component can then be inferred. If the component is unstable then the suggestion should be made to redesign to either increase the base area or lower the vertical c.g. position.
- *Tolerances.* By defining the tolerances attached to each surface of a component, an indication may be gained whether the surface is functional or not. A high tolerance probably indicates a precision surface with a high surface finish. If the tolerance is low then the surface is likely to be nonfunctional. If possible an analysis should be performed to check the validity of all the specified tolerances. Should a tolerance be found that is too tight then the designer should be asked whether the tolerance may be relaxed.
- *Surfaces that transmit force.* When modelling the assembly kinematics in the CAD modeller it is necessary to specify how each component moves relative to another. It should therefore be possible to determine which surface of a component transmits force to another component. If a designer attempts to modify this surface then he should be informed that the surfaces are already functional.
- *Surfaces that have motion.* From the CAD modeller it should be possible to identify surfaces that have relative motion between each other. If not, the parts are appropriate for part redundancy analysis.
- *Surfaces that provide location.* Any pair of mating surfaces can be provided with location features to facilitate easier alignment if no relative motion exists between the components. Such location features include pairs of: boss and hole, rib and slot and step and step. They form matching pairs on the two components to be joined. If the mating surfaces have a tight tolerance, or no reciprocal features or no functional surfaces are defined, then the suggestion should be made to provide location features.

- *Features that aid handling.* Features such as knurl, holes, boss and straight through slots aid handling and are therefore functional. This should be highlighted to the designer should he or she attempt to modify the feature.
- *Fastening features.* If a fastening feature is detected or specified then this part is theoretically redundant. A prompt should be raised to ask whether this part is absolutely necessary.

**(iii) Part redundancy analysis**

**Rules for part redundancy analysis**

1. If part needs to move relative to its mating parts during the normal function of the final assembly then do not combine. (If the motion is small then parts may still be combined.)
2. If it is essential that the part or subassembly be of a different material then do not combine. Advice can be provided at this stage to re-engineer the product to remove or change materials that could provide handling or feeding problems.
3. If the part or subassembly must be separated for maintenance or replacement then do not combine.
4. If the joining of the part or subassembly would prevent the assembly of another component then do not combine.

**Part redundancy data requirements** The following inputs are required by the module and have been determined from the above set of rules.

| Input | Provided by |
| --- | --- |
| Touching components | CAD |
| Relative motion | CAD, user |
| Material attribute | User |
| Component's functionality | User, functionality module |
| Separate for maintenance or replacement | User, functionality module |
| Prevent assembly of another component | Assembly planner, user |

- *Touching components.* Before components can be merged to form one, it must be determined whether these components touch each other in the existing design.

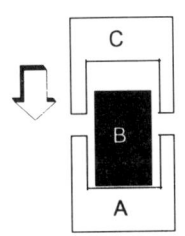

**Figure 4.6** Preventing assembly of another component

- *Relative motion.* Once two touching components have been identified, it must be determined whether any relative motion exists between these components. Should a large relative motion exist then the components cannot be combined. Smaller motions may be accommodated, for example by using an elastic hinge.
- *Material attribute.* If the two touching components are not made of the same material then the question should be asked "Can part A and part B be made of the same material?"
- *Component's functionality.* From a user input, the component's intended function can be detected. If the component is, for example a screw then the component's function is a fastener. As such this component is possibly redundant. Placards also serve no use in the functional operation of the product. However, it may be necessary to keep these within the product design for ergonomic or safety reasons.
- *Separate for maintenance or replacement.* If the components need to be separated for these reasons then they cannot be merged. A possible decision may be deduced from the component's functionality. For example a coil within an electrical assembly may require replacement during the product's life.
- *Prevent assembly of another component.* Where access to assembly is needed by another part (see Figure 4.6). If components A and C were merged to form one component then it would be impossible to assemble component B.

**(iv) Standard parts analysis**

**Rules for standard parts analysis**

1. If a suitable part exists in a standard component library then do not design a new component.

**Standard parts analysis data requirements** The following input is required by the module and has been determined from the above-mentioned rule.

| Input | Provided by |
| --- | --- |
| Component functionality | User |

- *Component functionality.* Before a new component is designed a check should be made to see whether a component already exists within the company's parts stock that has the same functionality. If so, then this should be highlighted to the user. This ensures that only essential "new" components are designed and manufactured.

### 4.4.3 Handling Analysis

Parts handling falls into three categories, manual, automated and robotic. In all three cases advice can be given on how the design may be improved by utilising non-functional surfaces and providing features that will aid the handling of that component.

When the functionality permits, components that are flexible or brittle should be avoided as these may be difficult to handle or feed. The material that the component is manufactured from should be carefully selected so that the component cannot be damaged during handling.

To reduce the amount of reorientation of components, they should be designed so that they are completely symmetrical. If this is not possible, then the asymmetry should be exaggerated to aid the orientation. If necessary lugs, notches, holes, etc. to aid handling should be included.

The handling module analyses components and assemblies to check their suitability for either manual, robotic or automated handling by considering such factors as geometry, material and weight.

**(i) Part handling**

1. Determine the weight of the part ⇒ The equipment database (see Section 6.4.2.(ii)) is interrogated for a suitable piece of equipment. If no equipment is suitable then the designer should be advised as such. If suitable equipment is available the handling rules applicable to the specific equipment can be applied.

Product Design for Assembly

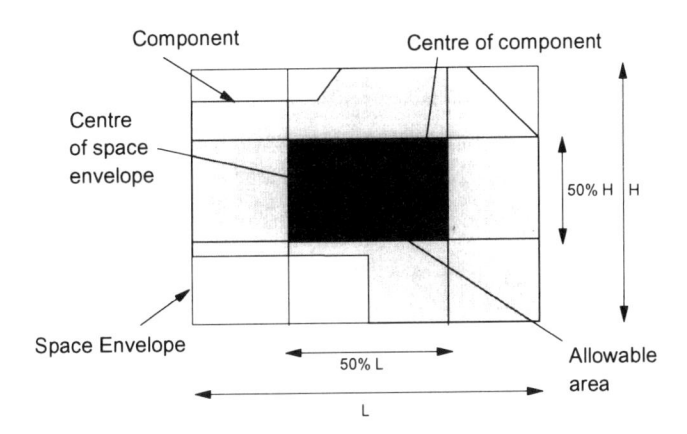

**Figure 4.7** Position for stable centre of gravity

2. Examine the part for handling features ⇒ If handling features have been placed on the component then go to item 6, else allow designer to add features or proceed to the next item.
3. Has the part got flat sides? ⇒ The flat surface of the largest area that is facing in the correct orientation such that it does not interfere with assembly operation should be highlighted. Otherwise the user should note that the assembly is difficult to handle as there are no dedicated handling features and no flat surfaces to grip the assembly.
4. Has the part got two flat parallel sides? ⇒ If there are flat surfaces then check whether any two surfaces are parallel. These surfaces should be highlighted to allow the user to select a pair as gripping surfaces. If not then advise the designer that handling would be made easier if there were two flat parallel sides.
5. Have either of these surfaces got a high tolerance? ⇒ If the answer is yes then the designer should be advised that the correct fixture must be used.
6. If the two sides or the identified handling features are used to handle the part then what is the c.g. of the part in this orientation? ⇒ If the c.g. is at an extreme end of the part then the user should be advised that the part could be unstable in this orientation. Figure 4.7 illustrates the method of determining a valid c.g. position. This is true for all the three principal axes.

|   |   |
|---|---|
| | 7. What is the parts symmetry? ⇒ If the component is almost asymmetric or symmetric the designer should be advised that if possible it should be made completely symmetric or asymmetric. |
| **(ii) Assembly handling** | 1. Determine the weight ⇒ Interrogate the equipment database for a suitable piece of equipment. If no equipment is suitable then the designer should be advised. Suitable equipment should be suggested to the designer. If it is accepted then the handling rules applicable to the specific equipment can be applied.<br>2. Examine the assembly for handling features ⇒ If handling features are present on the external surfaces of the assembly then go to item 6, else proceed to the next item.<br>3. Has the assembly got flat sides? ⇒ The flat surface of the largest area that is facing in the correct orientation such that it does not interfere with the assembly operation should be highlighted else the user advised that the assembly is difficult to handle as there are no dedicated handling features and no flat surfaces to grip the assembly.<br>4. Has the assembly got two flat parallel sides? ⇒ If there are flat surfaces then check whether any two surfaces are parallel and of a suitable area and orientation. If not then advise the designer that handling would be made easier if there were two flat parallel sides.<br>5. Have either of these surfaces got a high tolerance? ⇒ If the answer is yes then advise the designer that the correct fixture must be used.<br>6. If the two sides or the identified handling features are used to handle the assembly, what is the c.g. of the assembly in this orientation? ⇒ If the c.g. is at an extreme end of the assembly (See Figure 4.5) then advise the user that the assembly could be unstable in this orientation. |
| **(iii) Rules for handling analysis** | 1. If soft "fabric-like" and flexible components or brittle and fragile components are present in the assembly then try to engineer them out wherever possible or to place the part as late as possible in the assembly order.<br>2. If materials have the following characteristics: moisture sensitivity, static electricity, magnetic properties, then try to engineer them out wherever possible or to place the part as late as possible in the assembly order. |

# Product Design for Assembly

3. If the materials cannot be engineered out then the user should be advised to try and place them at the end of the assembly sequence. This is flagged to the assembly planner.
4. If the weight of the component is minimised then handling is made easier.
5. The following is an example of general vacuum gripper information:

Inner sealing lip:

| Area (mm$^2$) | Mass capable of being held (kg) | Area/Mass (mm$^2$/kg) |
| --- | --- | --- |
| 250 | 1.8 | 140 |
| 850 | 6.4 | 130 |
| 1960 | 14.7 | 130 |
| 3520 | 26.4 | 130 |
| 7850 | 58.9 | 130 |
| 17660 | 132.5 | 130 |
| 31400 | 235.5 | 130 |

Outer sealing lip:

| Area (mm$^2$) | Mass capable of being held (kg) | Area/Mass (mm$^2$/kg) |
| --- | --- | --- |
| 280 | 2.1 | 130 |
| 1100 | 8.3 | 130 |
| 2730 | 20.5 | 130 |
| 5080 | 38.1 | 130 |
| 10200 | 76.5 | 130 |
| 24870 | 186.5 | 130 |
| 42980 | 322.4 | 130 |

This leads to the following, general rule.

- If Area to Mass ratio of a component is lower than 130 mm$^2$/kg then vacuum gripping is unsuitable.

Minimum radius of curvature of the suction surface ($R_{min}$) that is acceptable for the given vacuum cup diameter:

| Diameter (mm) | Radius ($R_{min}$) (mm) | Diameter/$R_{min}$ |
|---|---|---|
| 18 | 50 | 0.36 |
| 33 | 110 | 0.30 |
| 50 | 165 | 0.30 |
| 67 | 165 | 0.41 |
| 100 | 306 | 0.33 |
| 150 | 465 | 0.32 |
| 200 | 758 | 0.26 |
| | | Average = 0.33 |

This leads to the following, general rule.

- If the diameter to radius of curvature ratio is greater than 0.33 then vacuum gripping is unsuitable.

6. The following is an example of general magnetic gripper information:

| Diameter (mm²/kg) | Area (mm²) | Mass capable of being held (kg) | Area/Mass (mm²/kg) |
|---|---|---|---|
| 6 | 28 | 0.4 | 70 |
| 13 | 133 | 1.9 | 70 |
| 19 | 284 | 2.8 | 101 |
| 25 | 491 | 7.5 | 65 |
| 32 | 804 | 10.7 | 75 |
| 51 | 2043 | 26.7 | 77 |
| 64 | 3217 | 35.5 | 91 |
| 76 | 4536 | 40.0 | 113 |
| 80 | 4998 | 177.6 | 28 |
| 86 | 5814 | 82.1 | 71 |
| 96 | 7296 | 99.0 | 73 |
| 100 | 7884 | 199.7 | 39 |

This leads to the following general rule.

- If the area to weight ratio for a component is lower than 28mm²/kg then magnetic gripping is unsuitable.

7. The following general algorithms can be used to calculate the holding force required:

**Vacuum grippers:**

| | |
|---|---|
| Gripping part top surface centre of gravity on axis of cup | Holding force = $3Mg$ |
| Gripping part top surface centre of gravity not on axis of cup | Holding force = $3Mg + 3MgX/0.633R$ |
| Gripping part side surface | Holding force = $3Mg/\mu$ |

**Magnetic grippers:**

| | |
|---|---|
| Gripping part top surface centre of gravity not on axis of magnet | Holding force = $3Mg + 3MgX/0.633R$ |
| Gripping part side surface | Holding force = $3Mg/\mu$ |

**Gripping cylindrical parts:**

| | |
|---|---|
| Centre of gravity between jaws | Holding force = $3Mg/\mu$ |
| Centre of gravity offset from jaws | Holding force = $3Mg(X - CP/2)/CP + 3Mg$ |
| Rotating part between jaws | Holding force = $6MgX/\mu D$ |

**Parallel grippers:**

| | |
|---|---|
| Centre of gravity between jaws | Holding force = $3Mg/\mu$ |
| Centre of gravity offset from jaws | Holding force = $3Mg(X - CP/2)/CP + 3Mg$ |
| Rotating of part between jaws | Holding force = $3Mg(X + CP/2)/\mu CP$ |

Where:
$M$ = mass (kg)
$g = 9.8$ m/s$^2$
$R$ = radius of vacuum cup/magnet (m)
$\mu$ = coefficient of friction
$X$ = distance of centre of gravity from jaws (m)
$CP$ = distance between contact points (m)
$D$ = diameter of cylinder or hole (m)

8. If a reasonably large flat surface is present on a component and the surface is not a base surface or adjacent to a neighbouring component then this component is likely to be suitable for automated gripping.
9. If a flat surface is present, the material is ferrous and surface finish is high and the surface is not a base surface

or adjacent to a neighbouring component then a magnetic gripper may be suitable.
10. If a flat surface is present and the surface finish is high and the surface is not a base surface or adjacent to a neighbouring component then a vacuum gripper may be suitable.
11. If a component has a large side area and/or the vertical position of its c.g. is low then the component will be stable in this orientation.
12. If handling is occurring after a feeding operation then the c.g. rule should not be triggered. Should the user not know whether the component is to be handled or fed then both the handling and feeding analyses should be run.
13. If large, flat, parallel surfaces are present on a component and the surfaces are not base surfaces or adjacent to neighbouring components then this component is suitable for handling by a gripper.
14. If complete part symmetry is not possible then exaggerate asymmetric features to aid orientation devices. (If necessary provide lugs, notches, etc. to aid orientation.)
15. If surface is non-functional then utilise it to provide handling and gripping features or change the attributes attached to the surface, such as surface finish or tolerances.

**(iv) Handling analysis data requirements**

The following inputs are required by the module and have been determined from the above set of rules.

| Input | Provided by |
|---|---|
| Material attribute | User, part redundancy analysis |
| Flat surfaces | CAD |
| Area of flat surfaces | CAD |
| Parallel surfaces | CAD |
| Area of parallel surfaces | CAD |
| Weight | CAD |
| Side area & c.g. | CAD |
| Symmetry | CAD |
| Tolerance attributes | CAD |
| Handling features | CAD |

- *Material attribute.* When a material attribute is assigned to a component, if the material is one of the following, then it should be flagged as being difficult to handle, and the

designer asked to try and choose an alternative material. The poor materials for handling are: rubber, ceramics, glass and any materials that are soft and fabric-like, moisture sensitive or magnetic, or generate static electricity.
- *Flat surfaces.* For a part to be handled easily with automated equipment it should have a least one flat surface.
- *Area of flat surfaces.* If the area of the flat surface is too small then it will be unsuitable for automated handling without specialist equipment.
- *Parallel surfaces.* For a part to be easily handled with an automated gripper, it should have flat parallel sides.
- *Area of parallel surfaces.* If the area of the parallel surfaces is too small then they will be unsuitable for gripping without specialist equipment.
- *Weight.* If a part is heavy then it will require specialist equipment. A check should be made to see whether the correct equipment exists within the company. If not then the designer should be advised and asked to try and reduce the weight of the part.
- *Side area and c.g.* By finding the area of each flat side and the vertical c.g. position if the side was placed horizontal, the best side for placing the component can be determined so that it resists being knocked over. This should be suggested to the designer.
- *Symmetry.* Boothroyd and Dewhurst (1987) defined the $\alpha$ and $\beta$ symmetry of a component to indicate the symmetry of the component. If $(\alpha + \beta) = 0$ the component is completely symmetric, if $(\alpha + \beta) = 720$ then the component is completely asymmetric. If $0 \leq (\alpha + \beta) \leq 720$ then the component requires a specific orientation. If $(\alpha + \beta) \neq 0$ or $720$ then the suggestion should be made to try and redesign the component to make it completely symmetrical or asymmetrical respectively.
- *Tolerance attributes.* If the tolerance is high then it can be inferred that the surface finish is good and the surface will be suitable for a vacuum gripper. If the surface is rough then this sort of gripper will not be suitable. The designer should be made aware of this.
- *Handling features.* If handling features such as a knurl, holes, boss, straight through slot, hook are present on a surface then this will aid handling, especially with large components. If a surface is non-functional then this surface can have a handling feature attached to it.

### 4.4.4 Feeding Analysis

Advice should be given on how parts can be designed so that they may be magazine fed as this reduces orientation problems. The positional uncertainty of the components that are presented to the assembly equipment is also reduced, which results in a reduced machine downtime. If robotic assembly is used then the need for a sensing system on the robot gripper is minimised. Feeding also allows a continuous flow of components, which must be at the rate required by the cycle time of the assembly equipment.

Should magazine feeding not be possible then components that tangle, nest or shingle should be redesigned, otherwise they will both be prevented from being bowl fed. Manual assembly would be possible, but it would be time consuming and frustrating for the assembly workers.

By palletising the assembly the positional uncertainty is again reduced. The pallet also allows the assembly to be handled more easily, quicker and safely as the handling forces are taken by the pallet.

The feeding module checks a component or subassembly for its suitability for feeding by considering such factors as geometry, size and weight.

**(i) Part feeding**

1. Determine likely feeding method ⇒ Feeding can be split into two categories; magazine feeding and "No specific orientation" methods. This determines the set of rules that can be applied.
2. Determine the material of the component ⇒ If the material is "fabric-like", flexible or brittle, etc. it should be suggested to the designer to change the material.
3. Determine the weight from the material and the volume ⇒ Interrogate the equipment database (see Section 6.4.2(ii)) for a suitable piece of equipment. If no equipment is suitable then the designer should be advised. The suggested piece of equipment should be proposed to the designer. If it is accepted then a further set of feeding rules can be applied, applicable to the equipment.
4. Examine the part for feeding features ⇒ Have feeding features been placed on the component.
5. Ask the user the intended feeding orientation and direction ⇒ By placing two components together in the intended feeding orientation and direction, the geometric rules can be

# Product Design for Assembly

applied to check for nesting, shingling and wedging—depending upon the intended feeding method. Advice should be given if any of these are likely to occur. If conveyor feeding is likely then proceed to items 6,7,8,9,10 else go to item 11.

6. Has assembly got flat sides? ⇒ A flat surface may be used as a tracking surface.
7. Determine area of all flat surfaces ⇒ The largest surface will probably provide the best tracking surface.
8. Has this surface got a high tolerance? ⇒ If the answer is yes then advise the designer that care must be taken or ask the user if the next smaller surface should be used, return to the previous item.
9. Determine the c.g. of the part in this orientation ⇒ If the c.g. is at an extreme end of the part then advise the user that the part could be unstable in this orientation. Ask the user if the next smaller surface should be used, return to item 7.
10. Flat, large, low tolerance, c.g. ⇒ Once these criteria are met this surface should be labelled as a tracking surface. If all the flat surfaces are used then the user should note that the assembly is possibly unsuitable for conveyor feeding.
11. What is the parts symmetry? ⇒ If the component is almost asymmetric or symmetric advise that if possible it should be made completely symmetric or asymmetric.

**(ii) Assembly feeding**

1. Determine the likely feeding method ⇒ Feeding can be split into two categories; magazine feeding and "No specific orientation" methods. This determines the set of rules that should be applied. The user should be questioned on which rule category to apply. If the user is unsure then both analyses must be performed.
2. Determine the total weight from the materials and volume of individual components ⇒ Interrogate equipment database for a suitable piece of equipment. If no equipment is suitable then the designer should be advised to try and reduce the weight of the assembly. The suggested piece of equipment should be proposed to the designer. If it is accepted then a further set of feeding rules can be applied, applicable to the equipment.
3. Examine assembly for feeding features ⇒ If feeding features have been placed on the assembly then go to item 6 else proceed to the next item.

4. Ask the user the intended feeding orientation and direction ⇒ By placing two assemblies together in the intended feeding orientation and direction the geometric rules can be applied to check for nesting, shingling and wedging—depending upon the intended feeding method. Advice should be given if any of these is likely to occur.
5. Has assembly got flat sides ⇒ A flat surface may be used as a tracking surface.
6. Determine area of all flat surfaces ⇒ The largest surface will probably provide the best tracking surface.
7. Has this surface got a high tolerance? ⇒ If the answer is yes then advise the user that care must be taken or ask the user if the next smaller surface should be used, return to the previous item.
8. Determine the c.g. of the assembly in this orientation ⇒ If the c.g. is at an extreme end of the assembly then advise the user that the assembly could be unstable in this orientation. Ask the user if the next smaller surface should be used, return to item 6.
9. Flat, large, low tolerance, c.g. ⇒ Once these criteria are met this surface should be labeled as a tracking surface. If all the flat surfaces are used then the user should note that the assembly is possibly unsuitable for conveyor feeding.

**(iii) Rules for feeding analysis**

1. If soft "fabric-like" and flexible components or brittle and fragile components are present in the assembly then try to engineer them out wherever possible.

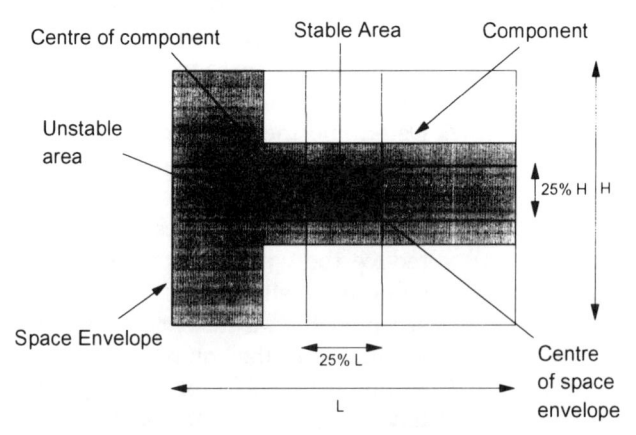

**Figure 4.8** Unstable centre of gravity position

# Product Design for Assembly

2. If materials have the following characteristics: Moisture sensitivity, static electricity, magnetic then try to engineer them out.
3. If the materials cannot be engineered out then the user should be advised to try and place them at the end of the assembly sequence. This is flagged to the assembly planner.
4. If the material is magnetic then the user is advised that a magnetic bowl feeder cannot be used.
5. If part can be magazine fed then there are no orientation problems. A check should be made for the likelihood of nesting and jamming.
6. If a part is likely to fall over while handling or feeding then try to position the centre of gravity of the part such that it either always tends to fall in the same orientation or assist in tracking. The centre of gravity of a component should fall within 25% of the centre of the three principal axes (see Figure 4.8). If the c.g. is at an extreme end of the part then the user is advised that the part could be unstable in this orientation.
7. If the part weight is reduced then it is easier to feed.

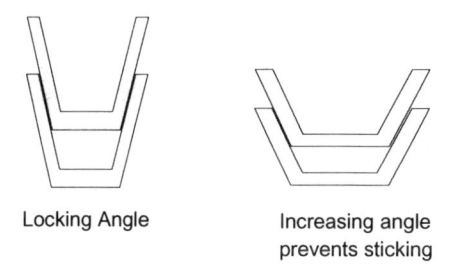

Locking Angle       Increasing angle
                    prevents sticking

**Figure 4.9**  Wedging parts (i)

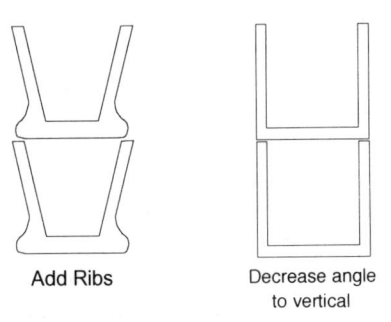

Add Ribs            Decrease angle
                    to vertical

**Figure 4.10**  Wedging parts (ii)

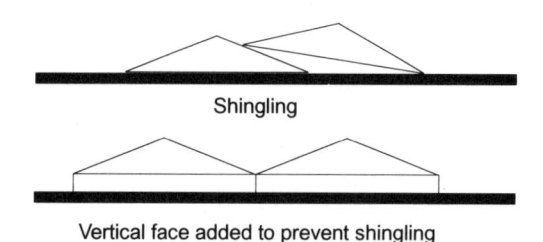

**Figure 4.11** Shingling parts

Shingling

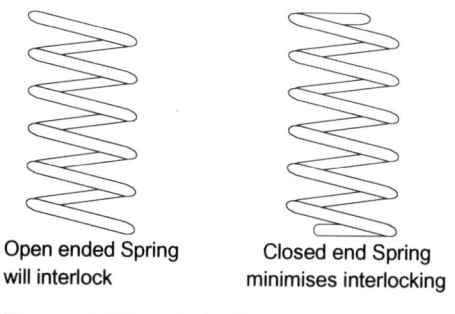

Vertical face added to prevent shingling

**Figure 4.12** Interlocking springs

Open ended Spring will interlock

Closed end Spring minimises interlocking

8. If the area of contact is small between the same two "open" components during magazine feeding then the tendency to wedge is reduced (see Figure 4.9, 4.10).
9. If the angle of the contact faces between the same two "open" components during magazine feeding is shallow then the tendency to wedge and engage is reduced.
10. If touching faces are vertical and contact area is large or the angle of each touching face is $45° \leq x \leq 90$ (a smaller contact area is required) then the tendency for parts to shingle is reduced.
11. If complete part symmetry is not possible then exaggerate asymmetric features to aid orientation devices. (If necessary provide lugs, notches, etc. to aid orientation.)
12. If part is non-rotational (a part whose basic shape is not a cylinder or regular prism whose cross-section is a regular polygon or triangular, or square parts that repeat their orientation when rotated about their principal axis through

angles of 120° or 90° respectively) then try to make the three principal dimensions differ by at least 10%.
13. If part attributes are present that encourage tangling, nesting or shingling then try to engineer them out (see Figure 4.11).
14. If the ends of springs are closed then the tendency to interlock is minimised (see Figure 4.12).
15. If the ratio of a component or assembly's weight to the volume of its envelope is less than 153 kg/m$^3$ then it is too light for a conventional hopper feeder.
16. If a component or assembly's smallest dimension is greater than 50 mm and its largest dimension greater than 150 mm then it is too large to be handled by a conventional vibratory hopper feeder.
17. If $L > d/8$ for a component or assembly where
    $L$ = length of the part measured parallel to the feeding direction
    $d$ = feeder or bowl diameter
    then it is too large to be fed by this particular feeder.
18. If a component or assembly's largest dimension is less than 3 mm then it is too small to be handled by a conventional vibratory hopper feeder.
19. General rules for vibratory bowl feeders are shown:

| Type of hopper | Component dimensions |
|---|---|
| Magnetic disc hopper | Almost any shape of component can be handled. Bulky components feed better than long slender rods. Inadvisable to operate with components having a length/width ratio exceeding 8:1, e.g. spark plug bodies |
| Side blade hopper | Spheres, discs, plain or headed pins, screws and bolts can be satisfactorily handled. The length/diameter ratio must exceed 2:1. Typical range—min. diameter 6 mm, max. diameter 38 mm, min. length 12 mm max. length 450 mm |
| Rocking through hopper | Mainly intended to handle cylindrical components with a length/diameter |

ratio of at least 15:1. Examples are the inter-stage handling of sewing needles during manufacture and the feeding of pins for commercial hinges

Centring hook hopper — Restricted to the handling of flat discs of metal, plastics, cork, card. A typical example is a cork insert for bottle caps 19 mm diameter × 1.5 mm thick.

Reciprocating funnel and tube hopper — Cylindrical components preferably having a length/diameter ratio not less than 1.5:1 or spherical components from 0.75 mm to 25 mm

Paddle wheel hopper — Discs or non-circular plates and other components having a length less than diameter or minor axis are the most suitable. Commonly used for such components as nuts, which may be circular, knurled, hexagon, square, etc., and rings, washers or small rectangular plates.

Horizontal funnel hopper — Capable of handling cylindrical components such as discs, cups or tubes having a length less than their diameter. It will also handle headed components such as flanged bushes, set screws etc., where the flange or head can be used to maintain control. An example is the feeding of an alarm clock hour wheel

Centreboard hopper — Spheres, plain or headed pins, screws, bolts and U-shaped components can be handled. Examples include 6 mm × 50 mm dowel pins and No. 8 Woodscrews × 38 mm

20. If parts are spherical or cylindrical, occupy a volume envelope in the order of 50 mm × 50 mm × 50 mm and their mass is less than 10 g then in general, they may be vibratory bowl fed.

## (iv) Feeding analysis data requirements

The following inputs are required by the module and have been determined from the above set of rules.

| Input | Provided by |
|---|---|
| Material attribute | User, part Redundancy analysis |
| Tangling, nesting and shingling attributes | User |
| Magazine fed (yes/no) | User |
| Intended feeding orientation | User |
| Contact area | CAD |
| Weight | CAD |
| Centre of gravity | CAD |
| Symmetry | CAD |

- *Tangling, nesting and shingling attributes.* If a spring is open then it is likely to tangle. If a component is to be magazine fed and the component is "open" then wedging is more likely if the contact area is large and the angle of the contacting surfaces is steep. If a component is not magazine fed then shingling is reduced if the angle of the touching surfaces lies in the range $45° \leq x \leq 90°$ and the contact area is large.
- *Magazine fed (yes/no).* If a component is not magazine fed then the components will not be preoriented before assembly. Alternative methods of feeding are vibratory bowl feeders and conveyors.
- *Intended feeding orientation.* The intended feeding orientation of the component must be specified by the user.
- *Contact area.* Once the orientation has been given it enables the contact surfaces of the component in the feeder to be inferred. The area of contact between the two adjacent components can then be found. If the area of contact is small ($\leq 50\%$ of total surface area of the touching sides) and the component is "open" then the tendency to wedge is small if the components are to be magazine fed. If the angle of the contact faces is shallow ($\leq 45°$ from the horizontal) then the tendency to wedge is reduced. If the touching faces are vertical and the contact area is large ($\leq 1\%$ of the total surface area of the component) or the angle of each touching face is $45° \leq x < 90°$ (a smaller contact area is required) then the tendency for parts to shingle is reduced. The designer should be advised if shingling appears likely.

- *Weight.* If a component is too heavy then it is unsuitable for vibratory bowl feeding. This should be highlighted to the designer.
- *Centre of gravity.* Determine the c.g. position of the component in 3D space. If the c.g. is not near the centre of the component envelope then the component is unstable. The designer should be asked whether the c.g. position will assist in feeding.

### 4.4.5 Positioning and Insertion Analysis

The positioning and insertion analysis checks the ease of positioning and insertion operations and provides advice on how the operation may be eased. Insertion is the most common method by which components are placed within an assembly. To aid insertion all "peg-in-hole" insertions should be from vertically above, using a single straight line insertion motion wherever possible. Coincidental insertions should be avoided. Generous lead-in chamfers should be provided. The widest permissible tolerances should be used. Tight tolerances result in extra costs, caused by additional operations, higher tooling costs, longer processing times, more scrap and rework, need for more skilled labour, greater materials costs and greater investment in precision equipment.

The positioning and insertion module checks on the ease by which two components can be inserted by considering such factors as their alignment, the presence of chamfers and tightness of tolerances.

#### (i) Analysis order

1. Identify components that are involved in the insertion operation.
2. Check alignment of components ⇒ If a collision is detected then advise the user.
3. Identify assembly direction ⇒ If assembly direction is anything other than vertical then user should be advised that if a peg and hole operation is to be performed then ideally it should be vertically downwards.
4. Identify coincidental insertions ⇒ Advise user that coincidental insertions should be avoided if possible by either changing the assembly sequence or redesigning the component. Should the user wish to continue with the analysis then the particular insertion to be analysed must be identified.
5. Identify whether any special alignment is necessary.

# Product Design for Assembly

6. Identify chamfers.
7. Identify tolerances.
8. Identify relieving on peg ⇒ If a peg and hole insertion is to be performed the user should be advised that the tendency to wedge is reduced if the peg is relieved.
9. Is a further operation required.
10. Identify the need for any special insertion techniques.
11. Determine assembly force.
12. Check equipment database.
13. Determine the overall probability of insertion ⇒ Should the probability of insertion be low then advise that the tolerances should be relaxed or a better chamfer added to both the "peg" and the "hole". Slightly convex fillets are more efficient otherwise, steep chamfers are preferable to shallow ones.
14. Free space ⇒ The insertion operation can be visually simulated by the user should the insertion operation be in doubt.

**(ii) Rules for positioning and insertion analysis**

1. If an insertion is to be performed then provide generous lead-in chamfers on both the two reciprocal parts. Slightly convex fillets are more efficient otherwise, steep chamfers are preferable to shallow ones (see Figure 4.13)
2. If an insertion is to be performed and it is not intended to be a push fit then try to relax the tolerances.
3. If an insertion is to be performed then it should be from vertically above using a single straight line insertion.
4. If an insertion is to be performed then the two reciprocal components must be aligned.

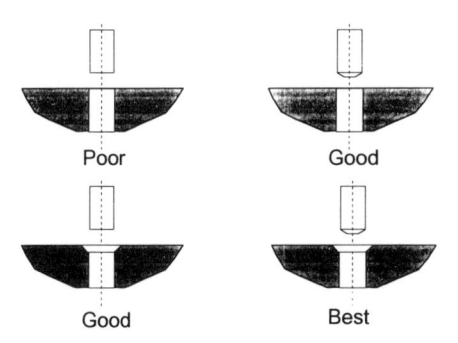

**Figure 4.13** Ease of insertion

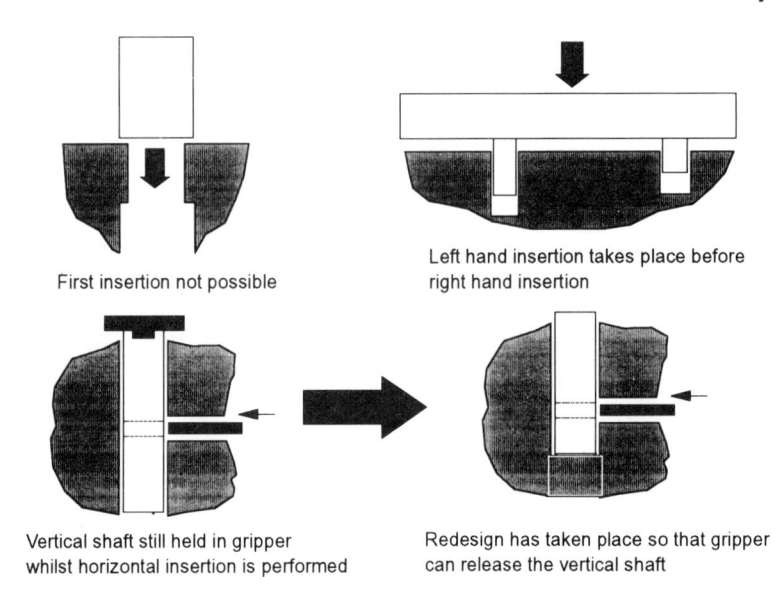

**Figure 4.14** Co-incidental insertions

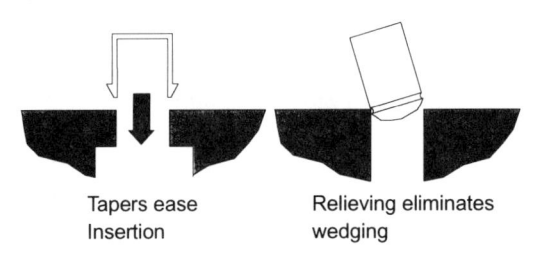

**Figure 4.15** Other insertion considerations

5. If location features, that require an insertion operation, are present to aid the alignment of components then these should be checked for their ease of insertion as per Rule1.
6. If a coincidental insertion is to take place then redesign the component such that one insertion takes place before the other.
7. If an in-line coincidental insertion is to take place then ensure that all the insertions are possible (see Figure 4.14).
6. If peg is relieved then the tendency to wedge in the hole is reduced (see Figure 4.15).
7. If components are placed together in a restricted assembly area then ensure that the outside components are placed in the assembly first.

# Product Design for Assembly

8. Standard insertion features include:
   Boss and through/blind hole
   rib and slot
   pad and pocket.

**(iii) Feeding analysis data requirements**

The following inputs are required by the module and have been determined from the above set of rules.

| Input | Provided by |
|---|---|
| Assembly directions | Assembly planner |
| Feature and mating feature | CAD |
| Chamfer feature | CAD |
| Tolerance attributes | CAD |
| Relieving on peg | CAD |
| Assembly force | CAD, FEA |
| Equipment | Equipment database |
| Additional operation required | CAD |
| Free space | CAD, Simulation |

- *Assembly directions.* From the assembly planner the assembly sequence or else the precedence constraints for assembly are available. From these it may be possible to determine the assembly direction of two particular components. From this information a check can be made to see whether relating features exist where an insertion type operation is performed.
- *Feature and mating feature.* Check for features and the reciprocal feature in the mating part. Such features are boss and hole, rib and slot and all location features. These features must be aligned for the insertion to be possible.
- *Chamfer feature.* Check whether a chamfer feature exists on the peg, hole or both. If a chamfer feature is not present then it should be suggested that some are added.
- *Tolerance attributes.* Check the tolerances on the peg and hole. If the tolerances are tight then it should be suggested that the tolerances are relaxed if the functionality is unaffected.
- *Relieving on peg.* Check for a groove feature on the peg as this will prevent wedging. If relieving is not present then suggest to the designer that a groove feature is added.
- *Assembly force.* Check whether the insertion is an interference fit. If it is then try to use FEA to determine the order of the assembly force required.

- *Equipment.* From the equipment database used by the resource planner a check can be made to see whether the correct equipment is available within the company to perform the operation.
- *Additional operation required.* If an additional feature is present such as a thread feature or a circlip then an additional operation is required after the operation to "lock" the assembly. This is obviously more time consuming and this should be highlighted to the designer, with the suggestion that if possible he or she should be try and reduce the number of operations.
- *Free space.* By performing a geometric simulation then the ability for the manual assembly worker or assembly equipment to access the assembly area can be assessed. If a simulation is not possible then in an area of high component density the designer should be asked to check whether enough space is available.

### 4.4.6 Joining Analysis

The joining analysis checks that the method by which components are joined is the most economic that is suitable for the application, i.e. must the component be separable for maintenance or replacement? This module focuses on the most common joining methods highlighted in the user requirement study.

The primary objective is to provide an approach to the selection of the main process groups and systems based upon the analysis of assembly requirements and process characteristics. The rules and selection tables presented cover a wide range of technical processes. They form a template that industrial companies can use to develop their preferred methods and rules. Interested readers are referred to reference in the bibliography for detail explanations.

From the resource planner equipment database (see Section 6.4.2(ii)) specific equipment can then be selected that allows specific process rules to be applied. The selection is primarily influenced by the function and cost.

**(i) Analysis order**

1. Identify components that are to be joined.
2. Identify the joining faces.
3. Identify joining features or attributes, if any ⇒ If these are present then the method of joining has already been decided, go to item 8 and suggest an alternative method.

# Product Design for Assembly

4. Identify location features or attributes, if any ⇒ If none are present then suggest that some are added.
5. Determine whether joint is to be temporary or permanent.
6. Is it necessary for relative motion to exist between the two components?
7. Identify the material of the two parts
8. Suggest a joining method
9. Look up standard parts list and equipment database ⇒ highlight parts that fall into the particular category. If the joining method requires additional equipment then check the equipment database.
10. Free space ⇒ the insertion operation can be visually simulated by the user should the insertion operation be in doubt.

**(ii) Rules for joining analysis**

1. If joint is temporary then the available processes are: snap-fit, screwing, bolting, clipping, press-fit and wrapping.
2. If joint is permanent then the available processes are: welding, adhesives, riveting, interference-fit and soldering.
3. If joint contains its own location feature (i.e. a snap fit) then this joint is better than one without a built in location feature.
4. If standard fasteners are used and their number is minimised then this reduces assembly cost and time.
5. If production volume/rate is high then consider soldering, welding, screwing, snapping, clipping and riveting.
6. If strength requirement is high then consider soldering, welding, mechanical joining and adhesives.
7. If fluid impermeability is needed then consider soldering and adhesives.
8. If joint must be flexible then consider special forms of mechanical joining (i.e. with washers)
9. If joint must be disassembled regularly then consider screwing and bolting.
10. If joint access is restricted then press-fits, interference fits, snapping and clipping.
11. If the visual appearance of the joint is important then press and interference type joints are best.
12. If mechanical type joining methods are used and the visual appearance of the joint is important then the joint should be from vertically upwards, preferably from the base component.

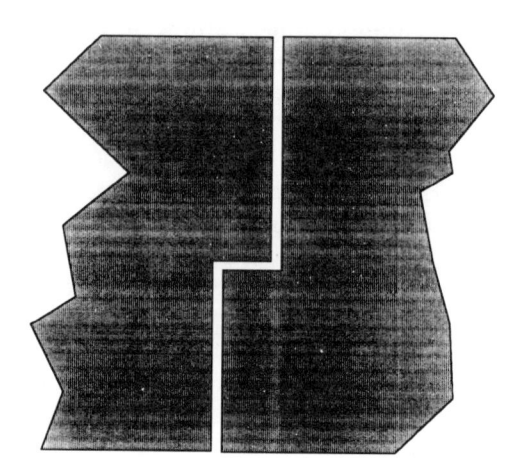

**Figure 4.16** Joint step

13. If a protective casing is to be joined then the joint line should be stepped to prevent inclusion of dust and foreign objects (see Figure 4.16).
14. If the area of contact between the two components is small then consider adhesives, soldering and welding.
15. If the material thickness is thin and a butt type joint is to be made then consider adhesives and soldering.
16. If the environmental factor of the joint must be considered then consider mechanical joining.
17. If standard components are being used then the joining method suitable for this component must be used.
18. If the material attribute of both components aids the selection of a joining technique then highlight this to the user.
19. If the material attribute of a component can be changed to use a different joining technique then this should be highlighted to the designer.
20. If components mate then consider the chemical interaction between materials.
21. If material is non-ferrous, stainless steel or non-metallic then corrosion properties are good.
22. If material attribute = superalloys then corrosion resistance at elevated temperatures is good.
23. If material attribute = gold, silver or platinum then corrosion resistance is good.
24. If material attribute = titanium then corrosion resistance at high temperatures is good.

25. If material attribute = ceramics then corrosion resistance is excellent.
26. If material attribute = graphite then chemical resistance is good.
27. If material attribute = stainless steel then chemical resistance is good.
28. If material is gold or silver plated then adhesives should not be used.
29. If material is lead then welding should not be used.
30. If material is an aluminium alloy then welding characteristics are poor and welding should be avoided.
31. If material is brass then welding characteristics are poor and welding should be avoided.
32. If material is a steel then all types of welding are suitable.
33. If material is a plastic then ultrasonic welding is suitable.
34. A simple morphology of typical assembly is presented below:

| Function | Possible methods | | | |
|---|---|---|---|---|
| Joining of sheet metal parts | Spot Weld | Rivet | Adhesive Bonding | Screw |
| Structural attachment to | Weld nut | Clinch nut | Stud weld | Adhesive bonding |
| Rotational retention | Circlip | Collar with or grub screw | Push on "Ratchet" ring | Nut and washer |
| Resistance to vibrational loosening | Stiff nut | Lock bolt | Anaerobic locking sealant | Slotted nut, pinned or wired |

35. A general comparison of joining process characteristics is presented in the following table:

| | Welding | Brazing and soldering | Mechanical fastening | Adhesive bonding |
|---|---|---|---|---|
| Permanent Joints | | Usually permanent (soldering may be non-permanent) | Threaded fasteners permit disassembly | Permanent |

|  | | Welding | Brazing and soldering | Mechanical fastening | Adhesive bonding |
|---|---|---|---|---|---|
| | | Local stress points in structure | Fairly good stress distribution | Points of high stress fasteners | Good uniform load distribution over joint area (except in peel) |
| | Joint Features | Generally limited to similar material groups | Some capability of joining dissimilar | Most forms and combinations of materials can be fastened | Ideal for joining most dissimilar materials |
| | | Very high temperature resistance | Temperature resistance limited by filler material | High temperature resistance | Poor resistance to elevated temperatures |
| P R O D U C T I O N | Joint Preparation | Little or none on thin material. Edge preparation for thick material | Pre-fluxing often required (except for special brazing processes) | Hole preparation and often tapping for threaded fasteners | Meticulous cleaning usually necessary |
| | Post Processing | Heat treatment sometimes necessary | Corrosive fluxes must be cleaned off | Usually no post-processing occasionally retightening in service | Some "post-assembly" time often necessary for full cure |
| A S P E C T S | Equipment | Relatively expensive, bulky and often requires heavy power supply | Manual equipment cheap. Special furnaces and automatic units expensive. | Relatively cheap, portable and can be used for "on-site" assembly. | Simple for manual- low-volume production. Dispensing and curing equipment for high production can be expensive. |
| | Consumable | Wires, rods, etc. fairly cheap | Some special brazing fillers expensive. Soft solders cheap | Quite expensive | Structural adhesives quite expensive |
| | Production Rate | Can be very fast | Automatic processes quite fast | Joint preparation and manual tightening slow. Mechanised tightening fairly rapid | Fairly slow particularly with chemical curing adhesives |
| | Quality Assurance | Non-destructive testing methods applicable to most processes | Inspection difficult particularly on soldered electrical joints | Reasonable confidence in torque control tightening | Inspection very difficult. NDT methods very limited. |

# Product Design for Assembly

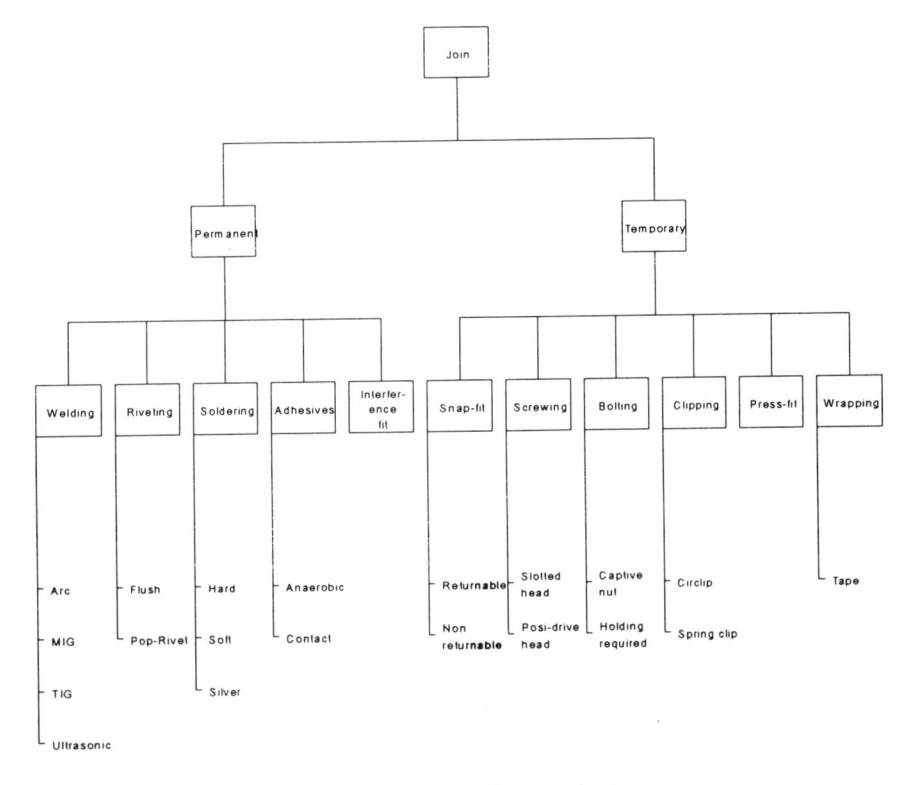

**Figure 4.17** Joining process selection chart

A joining process selection chart is illustrated (Figure 4.17), where a process and equipment selection could be worked out in a-step-by-step manner. Each user company will have a specific chart relating its product and process-specific selection procedure.

**(iii) Joining analysis data requirements**

The following inputs are required by the module and have been determined from the above set of rules.

| Input | Provided by |
|---|---|
| Touching components | CAD |
| Relative motion | CAD, user |
| Material attribute | User |
| Permanent or temporary join | User |
| Standard part | Standard part library |
| Fastening feature | CAD |
| User questions | User |

- *Touching components.* If a join is to be made then the two components must be touching.
- *Relative motion.* If there is relative motion between the two components then this will restrict the joining methods available.
- *Material attribute.* The material of the two components will affect the joining methods available.
- *Permanent or temporary join.* Whether the join is permanent or temporary will affect the joining method available.
- *Standard part library.* The company should have a library of standard parts, including catalogues from suppliers. If a joining method is selected that uses extra parts, such as non-standard nut and bolt, then a check should be made to ensure that if possible that these parts should come from the standard parts library.
- *Fastening feature.* A fastening feature may already have been attached to the component by the designer, or be an integral part of the component to be joined, i.e. a snap-fit. If this is the case the suitability of the feature to the design should be checked against the above criteria.
- *User questions.* The user can be questioned as to the type of joining requirements that are envisaged. From the questions an indication of the more preferable joining methods can be gained.

## 4.5 Conclusions

To summarise, the DFA module consists of five sub-modules, each of which has different data requirements and rule structures. The user has total control of the sequence of analysis through the multiple starting points of the analysis. With an integrated database, the amount of repetitive user input is minimised. As much information as possible is extracted from the CAD solid model and from other integrated modules. The sub-modules are designed so that they can analyse a design throughout the design process.

## 4.6 Bibliography

Ahm, T. and Fabricius, F. (1990) Integrating design for assembly with other design tools, *Conf. Proc. ICAA*, SME, Dearborn, MI.

Andreasen, M. M., Lund, T. and Kahler, S. (1988) *Design for Assembly*, IFS Publications.

Boothroyd, G., Dewhurst, P. and Knight, W. (1994) *Product Design for Manufacture and Assembly*, Marcel Dekker, New York.

Chal and Redford, A. (1994) *Design for Assembly*, Chapman & Hall, London.

Corbet, J., Dooner, M., Meleka, J. and Pym, C. (1991) *Design for Manufacture*, Addison Wesley, Reading, MA.

Kalpakjian, S. (1989) *Manufacturing Engineering and Technology*, Addison Wesley, Reading, MA.

Komacek, S. A., Lawson, A. E. and Horton, A. C. (1990) *Manufacturing Technology*, Delmar Publishers.

Kusiak, A. (ed.) (1992) *Intelligent design and manufacture*, Wiley, New York.

Niebel, B. W., Draper, A. B. and Wysk, R. A. (1989) *Modern Manufacturing Process Engineering*, McGraw-Hill, New York.

Society of Manufacturing Engineers (1993) *Tool And Manufacturing Engineers Handbook*, SME Publishing.

## 4.7 References

Atiyeh, P. G. (1992) Design for assembly: sometimes more is less, *Assembly Automation*, **12**(2).

Bedworth, D. D., Henderson, M. R. and Wolfe, P. M. (1991) *Computer-Integrated Design and Manufacturing*, McGraw-Hill, New York.

Boothroyd, G. and Dewhurst, P. (1987) *Product Design For Assembly Handbook*, Salford University Industrial Centre, Salford.

Holbrook, A. E. K. (1988) *Design Assistance For Complex Engineering Assemblies*, The CIM Institute, PhD Thesis.

Jackson, D. H. (1985) *Effects on Product Design of Components Assembled by Automatic/Robot Assembly*, The CIM Institute, MSc. Thesis.

Kim, G. J. and Bekey, G. A. (1990) A DFA analysis and advisor system, *Proc. 1990 IEEE Int. Conf. Systems, Man and Cybernetics*, 4–7 November, Los Angeles, CA.

Kim, G. J., Bekey, G. A. and Goldberg, K. Y. (1992) A shape metric for design for assembly, *Proc. 1992 IEEE Int. Conf. Robotics and Automation*, Nice, France, May.

Leaney, P. G. and Wittenberg, G. (1992) Design For Assembling, *Assembly Automation*, **12**(2), 8–17.

Li, R.-K. and Hwang, C.-L. (1992) A framework for automatic DFA system development, *Computers in Industrial Engineering*, **22**(4), 403–413.

Nieminen, J., Kanerva, J. and Mantyla, M. (1989) Feature-based design of joints, *International GI-IFIP Symposium*, Production Technology Centre, Berlin, November 8–10 1989.

SCOPES (1992) *Deliverable D3 User Requirements*.

Shi, J. (1991) *Computer Integrated Design For Assembly System*, The CIM Institute, MSc. Thesis.

Swift, K. G. and Doyle, R. (1987) Design for assembling in CAD using knowledge based systems in mechanical design, *Engineering Designer*, July/August.

# 5 Assembly Planning

## 5.1 Introduction

Assembly planning outlines the nature and the succession of operations necessary to assemble the product. An assembly plan describes how to assemble the product, i.e. specifies a sequence of assembly operations that has to be carried out in order to make the final product from its constituent parts.

From the point of view of the industrial companies, a sequence of operations is usually a tree (see Section 5.4.4(i)) whose leaves correspond to the parts of the product and nodes represent assembly operations, as shown on Figure 5.1.

This chapter describes the main steps of assembly planning:

- the assembly modelling, taking into account geometrical and technological data;
- the process modelling, which is the creation of the actions, linking processes and components;
- the creation of the assembly plans, based on manual or automatic procedures.

Referring to Chapter 3, assembly planning belongs to the off-line part of the CAD method and is mainly related to both the design and resource planning modules. These connections are shown in Figure 5.2.

## 5.2 Users' Needs

The requirements specified by industrial users can be summarised in the following points:

- user-friendly assembly sequence representation and edition
- automatic method;
- improvement of the module behaviour from user enrichment;

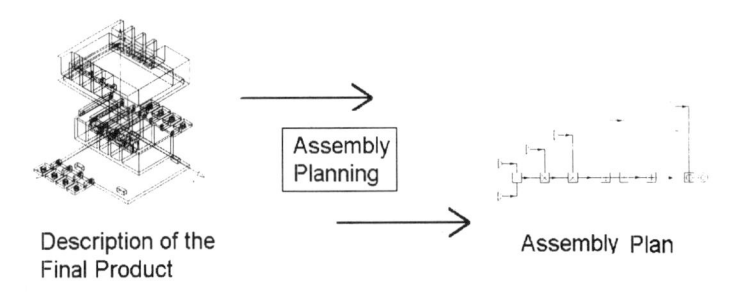

**Figure 5.1** A global view of the assembly planning

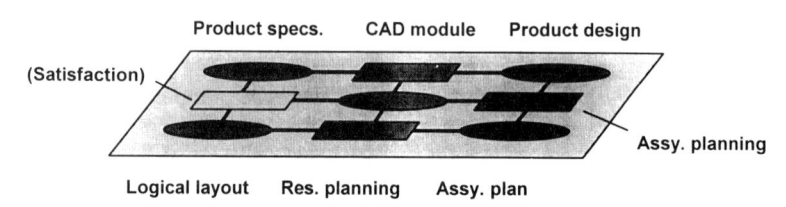

**Figure 5.2** Situation of the module within the CAD method for industrial assembly architecture

- management of subassemblies;
- possibility to take into account parallelism between operations.

Today, the process engineers usually draw an assembly sequence manually. They would appreciate having a dedicated computerised tool to speed up the edition through the representation and handling of the assembly sequences. This could be completed by the use of symbols corresponding to operations: insertion, marking, inspection, . . .

The assembly planning module has to be able to generate assembly sequences automatically because the industrial companies do not necessarily have previous experience of a similar product. In some cases, the process engineers start from scratch. This can help to speed up the generation of the sequence. The user must obviously be able to modify the result of the module.

The module should be able to integrate user knowledge on the product and its process, with criteria covering both design issues and manufacturing issues. This knowledge can be supplied by means of rules.

Engineers use rules that, from the point of view of an automatic module, can be seen as constraints that the module should try to comply with. Some rules apply to the parts themselves; some others may concern the operations. Finally, some rules integrate the impact of the assembly plan on corresponding technical solutions, and propose options for better manufacturing.

Major constraints for the assembly sequence consist of the definition of the subassemblies. Thus, the user must have the opportunity to split the product in predetermined subsystems.

According to the possible flexibility of an assembly line with respect to the control of the assembly sequence—permutation of the operations—assembly plans do not necessarily need to be made of sequences of fully ordered operations. The user may want to select at this stage between the two strategies of following sequential paths along a tree or allowing some degrees of freedom in the manufacturing process.

## 5.3 State of the Art

Most authors divide Assembly Planning into three steps, each one involving specific concepts:

- *Product modelling.* A distinction has to be made between the geometry of components and the description of relations between components in the final assembly.
- *Solving strategy.* This is the more divergent discussion point. The solving strategies described in the literature lead to either one plan only, or several plans, or the whole set of assembly plans. Many of them use heuristics, precedence relations and collision-free paths. All of them belong to the "graph search" family.
- *Selection and evaluation of assembly plans.* This is required if many assembly plans have been generated.

The literature on assembly planning is numerous and many different approaches are described in papers. The following sections lead to show both the common and the divergent points amongst these approaches, and also the missing features that we try to solve in the method presented in this book.

### 5.3.1 Product Modelling

In general, product databases described in the literature have the following features:

- either a free-form database, which can be considered as academic but comprising all needed features, or a CAD

database, industrial but incomplete for running the Assembly Planner;
- either purely geometric (Mascle, 1990, 1992; Mascle and Figour, 1990) or containing technological data;
- either with or without "grouping elements", such as bill of material (BOM) or clusters of parts, which hardly simplify the assembly planning problematic (Heemskerk *et al.*, 1990).

Some authors wish to extract automatically all needed data from a purely geometric CAD database. The problem is obviously difficult because data extraction depends on the structure of the geometric database, which is not standardised. Therefore the problem has not been deeply investigated. The use of parametrised and feature-based CAD systems introduces new and powerful data structures. At this time, we think that it is the "way to the future" for assembly planning. Such a system is especially interesting from the point of view of the relationship between components:

- *geometrical contacts* as face to face or cylindrical (mating features);
- but also *attachment type* as fastener or simple insertion (Homem de Mello and Sanderson, 1990, 1991a,b; Homem de Mello and Desai, 1990; Park *et al.*, 1991).

Many authors wishing to represent the concept of contacts between components, use a simple contact graph as face to face or cylindrical to cylindrical as in Wolter (1990, 1991).

Assembly planning needs to define relations between remote components. This can be described e.g. by *collision free paths* for removal a component (Santochi and Dini, 1992; Dini and Santochi, 1992) or by *precedence constraints* between parts (Park *et al.*, 1991).

Grouping *a priori* some components in a view to reduce the combinatorial problem is often used by authors (Heemskerk *et al*, 1990; Henrioud *et al.*, 1991; Mascle, 1990, 1992; Mascle and Figour, 1990; Zhang and Zhang, 1990). Two means for this are:

- *clustering parts*, i.e. considering several parts as one only, because the assembly order is not important e.g. a cluster of screws;
- *layering parts*, i.e. defining a hierarchical bill of material, containing subassemblies.

### 5.3.2 Solving Strategies and Assembly Plans Representation

In the literature, the main discussion points are related to the solving strategy. It would be too long to explain in detail each of these strategies. Therefore, we present only a typical approach, summarising the most frequently used ideas found in the literature.

An assembly planner can be described as using:

- a disassembly approach;
- a tree representation of the assembly plans;
- precedence constraints and technological heuristics—both to avoid spurious assembly sequences and to reduce the combinatorial problem.

It is important to say that authors have divergent ideas on both the ultimate goal of an assembly planner and the assembly plan representation. Some lead to generate one only plan (De Fazio et al., 1990; Baldwin et al., 1991; Wolter, 1990, 1991), others several plans (Henrioud et al., 1991; Delchambre, 1990a,b, 1992; Delchambre and Gaspart, 1992; Delchambre and Wafflard, 1990, 1991; Delchambre et al., 1991), and the rest the whole set of assembly plans (Wolter, 1990, 1991; Park, et al., 1991). Some work with sequences (Homem de Mello and Sanderson, 1990, 1991a,b; Homem de Mello and Desai, 1990; Baldwin et al., 1991; Wolter, 1990, 1991; De Fazio et al., 1990; Zhang and Zhang, 1990), other with trees (Henrioud et al., 1991; Wolter, 1990, 1991; Park et al., 1991), the rest with more general graphs (Miller and Stockman, 1990). These facts imply many variations in the methods. These can be very different depending on the aim of the author; e.g. a first author wishes to generate the whole set of assembly sequences, a second one wishes only to obtain the best assembly tree, a third one is interested by a restricted set of "good" assembly trees. In this field, each one has his own ideas.

One can say that all these methods have a common point: they use a "graph search" method, optionally mixed with an evaluation (and elimination) of plans considered as worse than other ones.

Nobody currently tends to consider meta-heuristic methods.

### 5.3.3 Evaluation and Choice of Assembly Plans

Among the solving strategies we have described, some are not enumerative and lead to generate only a restricted set of plans. In this case, evaluating and sorting plans is mandatory

(Baldwin et al.). This means that evaluation means have to be defined. They are closely related to the assembly technology. The following lines describe the evaluation means often found in the literature.

- The *number of reorientations* (assembly turnovers) required to assemble the product gives a good idea on the value of an assembly plan. A single direction is preferred to multiple directions of assembly. (Baldwin et al., 1991; Park et al., 1991; Henrioud et al., 1991; Wolter, 1990, 1991).
- The *parallelism* between operations indicates the ability of the plan to allow as many degrees of freedom as possible for scheduling according to the production means of the line. When a plan is highly parallel, there are many more ways to dispatch the resources than when it is sequential. Several evaluations of parallelism are described in the literature, e.g. the number of components in the two subassemblies giving an assembly operation (Homem de Mello and Sanderson, 1990, 1991a,b; Homem de Mello and Desai, 1990; Miller et al., 1990; Park et al., 1991).
- Evaluating the *subassembly stability* is very interesting. It avoids mating parts which tend to disassemble spontaneously under the action of gravity or vibrations. The problem of this evaluator is the computation time, if the user wishes to be accurate (close to the geometry of the product), it can be very expensive. Furthermore, the laws of vibrations imply many complementary computations. Some ways exist that are not too complicated to be implemented, e.g. an evaluation of the degrees of freedom of the components involved in an assembly operation (Santochi and Dini, 1992).
- The literature also describes the *operative complexity* as evaluation mean of an assembly plan. The basic idea is that a subassembly subject to reorientation and/or translation should be easily manipulable by tools or hands. Therefore, the manipulability is the efficiency in orienting and in handling. This is linked to size, shape and weight of the involved subassembly (Henrioud et al., 1991; Wolter, 1990, 1991).
- The *cost of the equipment* or the *duration* of the operations, when they are known, are other popular evaluation means. (Heemskerk et al., 1990; Santochi and Dini, 1992).

# Assembly Planning

## 5.3.4 Pending Issues in the Literature

After this synopsis of the various product models, resolution strategies and assembly plans evaluation, the following unsolved points can be drawn:

- Data extraction from a geometric CAD system remains an unsolved problem, but it takes currently less importance due to the increased use of parametric features in CAD.
- Few papers showed an efficient (i.e. not time-consuming) way to evaluate the stability of a subassembly within an industrial CAD system.
- The manner to select assembly plans (after evaluation) is not clearly defined.
- The assembly planning applied to part families is a problem not currently developed in the literature, only for some particular case studies (Eskicioglu, 1990; Barnett, 1990).
- The influence of the tolerances on the generation of assembly plans is currently not developed in the literature. We found in Requicha (1989) a model for representing tolerances within a geometrical database of an assembly.
- The user-interface design and the learning of the assembly planning concepts to non-experienced users (hard problem!) is never taken into account.

## 5.4 Functionalities and Methodology

### 5.4.1 Introduction

The assembly planner can be divided into three steps:

1. assembly modelling;
2. generation of actions and constraints;
3. creation of assembly plans.

This last one can be defined in three scenarios:

1. a manual scenario;
2. a semi-automatic scenario;
3. an automatic scenario.

Figure 5.3 shows a global view of the architecture of the proposed assembly planner.

Let us describe the major functionalities of the assembly planner.

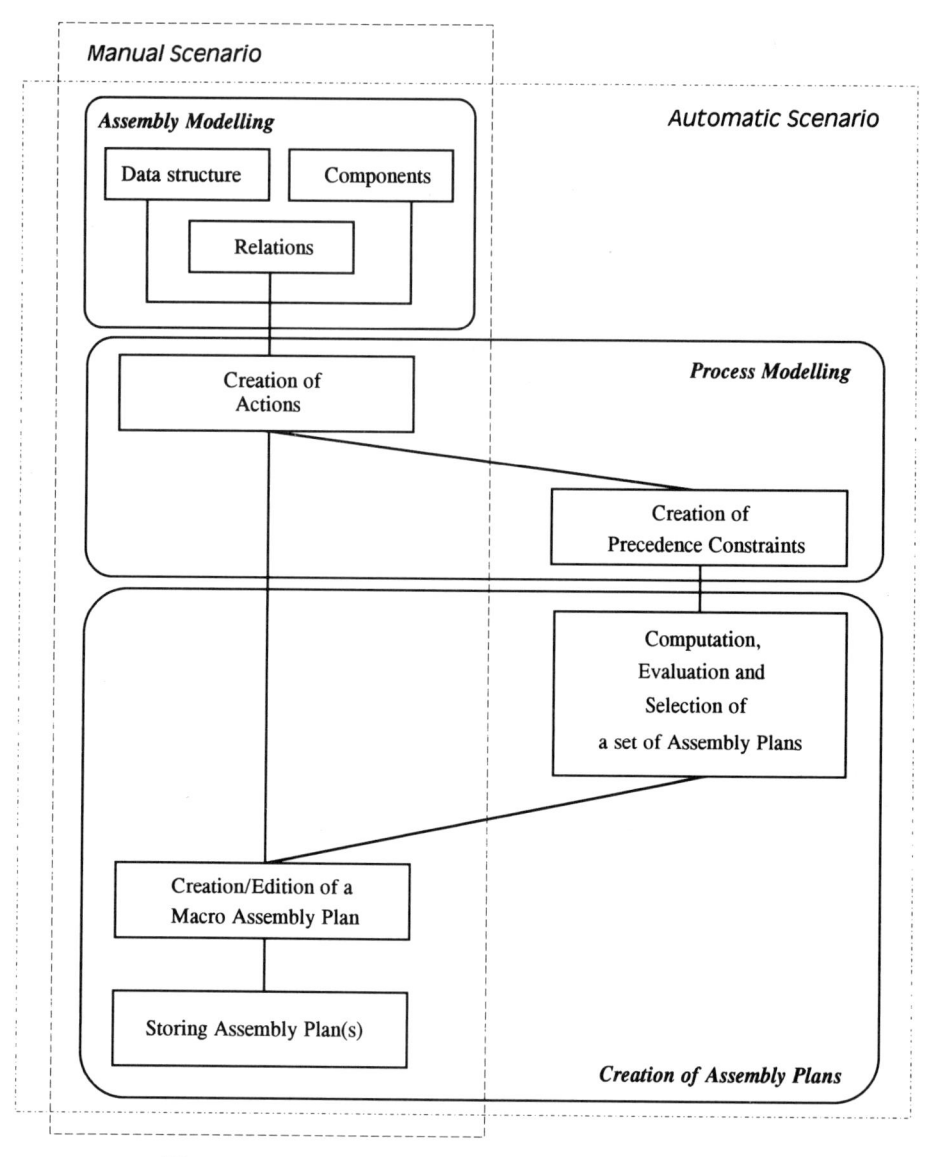

**Figure 5.3** The Assembly Planner Architecture

### 5.4.2 Assembly Modelling

We suppose that the product for which the assembly plan will be computed has already been modelled in a CAD database. Usually some additional data, necessary to carry on the assembly plan generation, are missing. These data are described in this section.

## (i) The product structure

We have already explained in a previous section that a product structure can be represented by a bill of material (BOM). These data are specified through the man–machine interface.

- *Determination of base parts.* In an assembly line, a component (generally presenting minimum handling) can be identified as a base component. This part, mounted on a transfer pallet, will move through each station of the line, receiving the other various components of the assembly. This optional information is currently provided by the operator.
- *Determination of clusters.* We define a cluster as a set of components to be assembled in the same operation and for which no assembly order is to be defined. For instance, a cluster can be three similar screwing required to secure a cover. The aim of this concept is both to simplify the representation of the product and its assembly plans and to reduce the combinatorial degree. Each cluster will be considered as a single component. Currently, clusters are specified by the operator, by specifying the involved components.
- *Determination of subassemblies.* The idea is to determine *a priori* some subassemblies of the final product in order to do some checks on the created assembly plan and to decrease the combinatorial degree in the automatic generation of assembly plans. Instead of working on the final product, the method is applied to two smaller subproducts. The operator specifies such subassemblies.

## (ii) The component features

The aim of this section is to describe the functionalities related to specific components of the BOM. The component features described in this section are the functional and the qualitative attributes. These attributes are specified through the man–machine interface. The geometrical data are defined in the CAD modeller.

- The *functional* attributes allows the user to define a component as a standard part, requiring particular assembly sequences, or deformation during assembly process, etc. or to which specific rules are attached. The current list of standard parts is screw, nut, bolt, nail, rivet, clip, seal and pin. These standard parts constitute one of the basic elements for the generation of assembly constraints, useful for computing assembly plans and checking edited assembly plans.

- The *qualitative* attributes allow definition of an item as fragile, inexpensive, and other qualities that will force the instanciated component to be assembled as soon or as late as possible. One can define the *latest components* (e.g. expensive, fragile, ...), which will be assembled as late as possible and the *earliest components* (e.g., inexpensive, easy to supply, ...), which will be assembled as soon as possible.

## (iii) The relation between components

A method to design an assembled product leads to consider physical contacts between components, called a *relation*. The two main aspects of a relation are the geometrical link (e.g. face to face) and the technological link (e.g. adhesive bonding). Both are stored in the model. Some other features have to be defined within the assembly planner, such as the assembly movement for establishing contact and the degree of freedom of a relation. As for the component design, the definition of the relations between components should be done in another module of the CAD modeller.

- The *geometrical relation attribute* is deduced from the physical contacts (degrees of freedom) between the components and is mandatory. We call it "insertion", "putting on", "sliding on", etc., depending on the number of degrees of freedom of the relation: from 1 (insertion) to 4 or 5 (putting on). The geometrical relation attribute is determined by geometric reasoning (Figure 5.4).
- The (optional) *technological relation attribute* completes the geometrical one. It is related to either a standard component or an additional process.
  —With each *standard component* or *fastener* is associated a set of rules and constraints. These will be applied to the com-

Insertion        2-dir insertion        Putting on

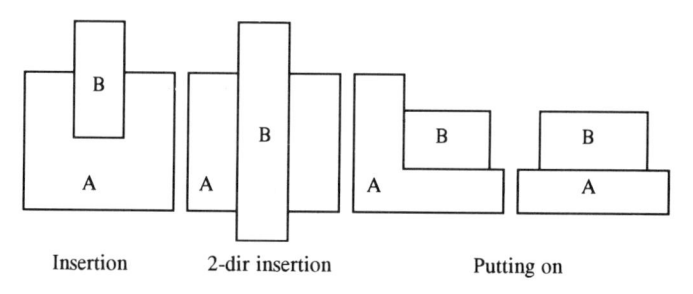

**Figure 5.4** Different types of geometric relations

ponents involved in the relation, in order to generate assembly constraints (see next section). Such rules have been defined for each standard component and fastener. For instance, a standard component "bolt" (= screw + nut) can imply two different set of rules: either "screwing bolt with the nut first" or "screwing bolt with the screw first". Another example could be screwing a screw within a set of stacked parts. In most frequent cases, the female part is placed first and below; in other cases it is placed latest, due to the geometry of the assembly. Each of these cases generates different set of rules and constraints.

This choice, done by the user, implies different rules and constraints on the assembly sequence.

— The *additional processes* are not linked to a standard component. But they have also, as standard components, implications on the assembly constraints. For instance, for "soldering" we consider "manual soldering" and "automatic soldering", which imply different rules and constraints on the assembly sequence. These attributes are specified by the operator from its technological knowledge. We have defined rules for the following processes: adhesive bonding, crimping, seaming, pressure fit, welding, manual and automatic soldering.

- The determination of the *assembly movement to establish contact* between the two components of the relation is a particular type of data. The complexity of such a movement can be

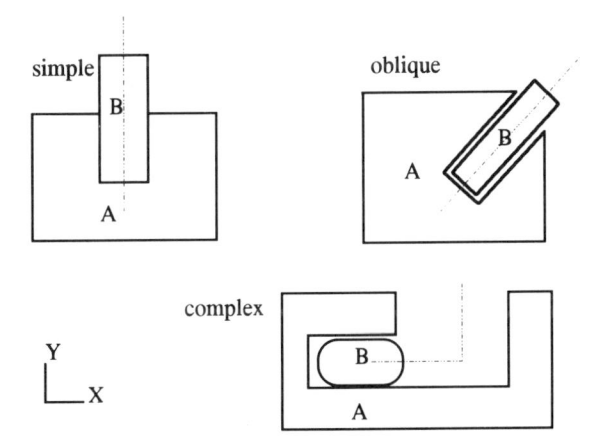

**Figure 5.5** Different assembly movements

sorted into three classes: *simple* (parallel to axis, in a straight line), *oblique* (not parallel to axis but still in a straight line) or *complex* (neither parallel to axis nor in one only straight line). We use this assembly movement for the creation of geometric precedence constraints. The simple and oblique movements are computed from degrees of freedom of the relations and by detecting collisions of a component moving a bit following each assembly direction. The complex ones are currently specified by the operator. The automatic generation of such assembly movements is currently investigated and effective methods (also called "motion planning methods") are now available but these are outside the scope of this book.

### 5.4.3 Process Modelling

The first goal of this section is to define the *actions* used in the assembly planner. The concept of "action" is used by most modules of this computer-aided method for industrial assembly. Its definition is enhanced in every following module. The result of this step is the creation of the "bill of actions", either manually or automatically created.

The second goal of this section is to define the constraints used in the automatic scenario. Their aim is to avoid the generation of spurious and numerous assembly plans. The result of this step is the creation of the "bill of constraints", mandatory for the semi- and fully automatic scenarios.

These data are related only to the processes and not to the product.

#### (i) The action

After specifying the assembly model, the operator (or the system) has to define the different actions involved in the assembly. This step is mandatory for all methods.

In general, an action can be defined as a positioning operation followed by an assembling operation. In other words, it describes the assembly technology, generally related to a relation (e.g. "screwing operation"), but not necessarily so. If the action is not to be linked to any particular relations (e.g. "painting opeation"), the user (or the system) must define the set of components to be processed with that particular action. An action itself can be an aggregation of actions; then we call it *macro-action*; it can be a part of another action, then we call it *micro-action*.

Sets of rules or algorithms are to be defined for each action.

For instance, screwing A requires all elements to be screwed by A to be already present and positioned.

At the assembly planning level, five families of actions can be defined:

- *feeding*, which is the initial step of the assembly plan and brings a component on the assembly line;
- *delivery*, which is the final step of the assembly plan;
- *operation*, which is an assembly operation or another type of operation, in other words, a modification of characteristics;
- *inspection*, coming from the user edition;
- *process*, which is an aggregation of actions; it is used to define a hierarchy of actions.

One should note that the action entity is used for almost each module of the design process, being then a fundamental element of the concurrent engineering approach (see also Section 3.2.1).

As noted, defining a hierarchy of actions is possible. We call every process instantiated with one (or several) component(s) an *assembly macro-action*. For instance:

◇ Process = SCREW Component
◇ Component = COVER no. 4376
⇒ Macro-Action = SCREW COVER no. 4376

From some relations, the automatic generation of macro-actions is possible. For instance, from a relation attribute "screw" between components "box" and "cover", the system can deduce the action "feed box, insert cover, screw cover". For other relations, it is not possible and the user has to specify the action.

The bill of actions (BOA) is the whole set of macro-actions, each component (or set of components) being instantiated to a process.

Another important functionality related to the concept of actions is to take into account "non-assembly" processes, i.e. *non-assembly macro-actions*, nevertheless mandatory in an assembly line. Examples are:

- inspecting (the quality of a process);
- painting (one or several components);
- printing (the brand name).

In these cases, an instance of a macro-action is:

◇ Process = PAINT List of Components
◇ List of Components = { LEFT_CAR_DOOR, RIGHT_CAR_DOOR }
⇒ MacroAction = PAINT { LEFT_CAR_DOOR, RIGHT_CAR_DOOR }

Both resource planner and simulation require details missing in macro-actions. So we can define a micro-action as a part of a macro-action. In other words, a macro-action is composed of a set of micro-actions.

For instance, consider the macro-action "SCREW A AND C" (Figure 5.6). We suppose that the operator chooses A as base part. Therefore the macro-action splits automatically into "FEED A", "SLIDE C ON", "SCREW". If a sufficient knowledge on the process is embedded in the system, it can be "exploded", e.g. as shown in Figure 5.7.

Exploding a macro-action into micro-actions requires that additional data be supplied to the system, e.g.. number of washers, screws involved, etc. We have defined the micro-actions needed for each standard component or fastener and for each additional process. These parameters are specified by the user and progressively fill the integrated database.

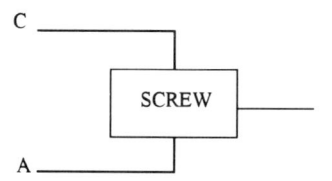

**Figure 5.6** An example of a macro-action

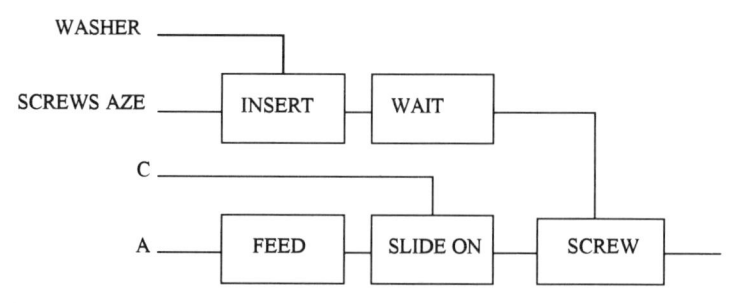

**Figure 5.7** An example of micro-actions

## (ii) The precedence constraints

A precedence constraint can be defined for any type of action and is generally defined by a precedence link between two actions. Precedence constraints are required to assemble or secure the complete product and are related to:

- the *geometry* of each component of the final assembly—its analysis leads to the generation of the "obstacle database", describe below;
- the different *processes*—the associated technological rules lead to the generation of the technological constraints;
- the *operator* knowledge—the operator may define individual constraints, coming from expertise on the product or the process.

These constraints represent the main input for the automatic scenario of the assembly plans generator, in order to avoid the generation of spurious and numerous assembly plans. The result of this step will be a list of constraints, also called "bill of constraints" (BOC).

The *geometrical precedence constraints* (Figure 5.8) define the constraints between positioning actions, by analysing the geometry of the components of the product. They will take into account the collisions occurring when trying to remove a particular component from a group of other components, along a specific direction.

For each component (e.g. A), the solid envelope of the geometrical definition of the component is computed. Along each direction (e.g. X+), this envelope is stretched as to reach the outside envelop of the product. On that path, parts in interference (e.g. B, C, D) are identified. The stored information is:

"*obstacles to removal of A in direction X are {B, C, D}*"

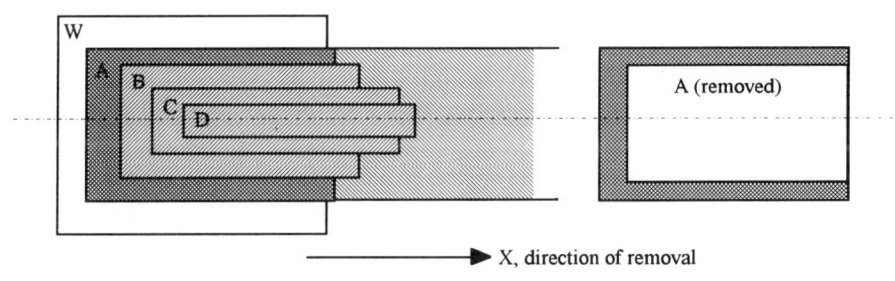

**Figure 5.8** An example of geometrical precedence constraint

which means that A, along direction X+, collides with B, C and D.

We automatically conclude that before being able to disassemble A from the group of parts in the direction X, B, C and D must be already removed from this group. In other words, A must be assembled before B; C and D, in the direction X–.

The *technological precedence constraints* are automatically generated from the standard processes applied to some components, as defined by the operator (see Section (i)). They lead to the best efficiency of the process. For instance, if the following feature is defined in the product "SCREWING A, B, C with SCREW S", the system can automatically generate these constraints:

- "Positioning A" must precede "Positioning B",
- "Positioning B" must precede "Positioning C",
- "Positioning C" must precede "Screwing SCREW S".

"X must precede Y" means that the left-hand action X has to precede the right-hand one Y within the assembly plan. In other words, this means that, after A is placed on the line, one may place B (on A) at any time. After B is placed on the line, one has to place C (on B). And so for screwing S: after having placed C.

The *user-defined precedence constraints* are defined because an expert system can not totally model the expert knowledge. Many constraints related to the other modules (a.o. choice of equipment) are not taken into account for the automatic creation of assembly plans. Therefore it is mandatory to give the operator the possibility to input manually other constraints to complete the BOC.

These constraints are given as "Action 1 must precede Action 2", Action1 and Action2 being specified by the user.

## 5.4.4 The Manual Procedure

The goal of this section is to describe the assembly plan creation, on the basis of the manual scenario. A user requirement is to use a tree to represent an assembly plan.

The manual scenario takes as input the BOA already defined in Section 5.4.3(i). The operator has to order these actions in order to build an assembly tree. This will be done by means of an assembly plan editor.

In general, an assembly tree containing only assembly pro-

# Assembly Planning

cesses is not ready to be sent to other modules because some data are missing. For this reason, other functionalities have been defined with a view to completing it.

**(i) The assembly tree representation**

Designing an assembly plan amounts to drawing an *oriented tree*. A tree is a type of graph, where:

- one node is designated as the root; in our case, it is the delivery of the final assembly;
- several nodes are designated as the leaves; in our case, they are the feeding of the components of the BOM;
- there is exactly one path between an arbitrary node and the root.

An example of an assembly plan is shown in Figure 5.9.

As already explained above, several levels of details can be used to represent the assembly operations (e.g. "SCREW A, B & C" or "SCREW A with V1" or "STACK A"), called micro-actions and macro-actions. An efficient editing procedure is to use these different levels of details simultaneously.

**(ii) Creation of an assembly plan**

The display is divided into three parts:

- the bill of material, i.e. the list of components;
- the library of actions, e.g. inserting, screwing, feeding, painting;
- a graphical window for drawing the assembly tree.

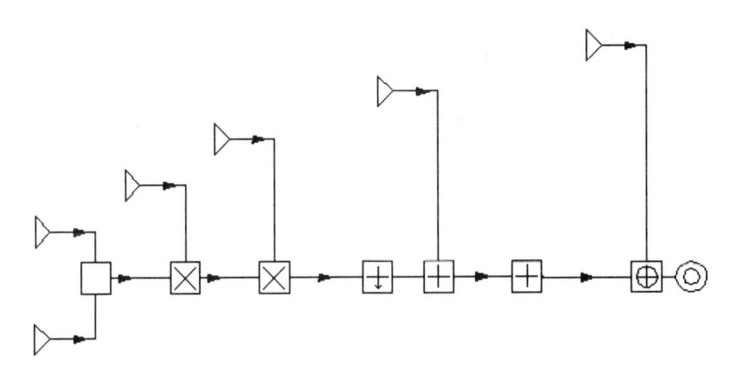

**Figure 5.9** An example of an assembly tree

| BOM: | Actions Lib | Assembly Plan: |
|---|---|---|
| E | Feeding | |
| D | Delivery | |
| C | Insertion | |
| | AddOp | |
| | etc. | |

**Figure 5.10** The manual creation of assembly plan at the initial step

The user fills the graphical window until every component of the BOM is associated while an action and every action is linked to another, forming a tree. Furthermore, the user completes the tree by supplying additional actions, here called "AddOp".

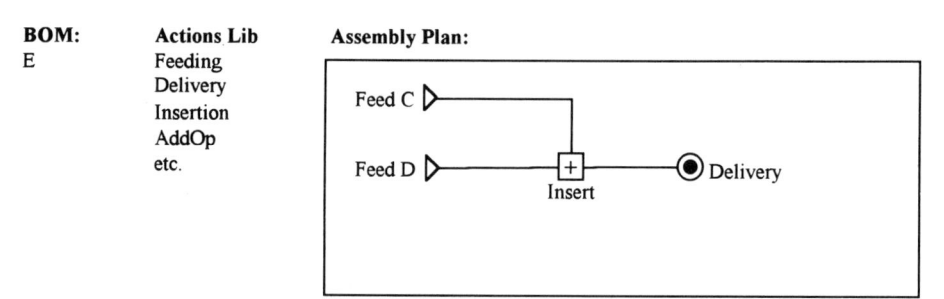

| BOM: | Actions Lib | Assembly Plan: |
|---|---|---|
| E | Feeding | Feed C ▷──────┐ |
| | Delivery | │ |
| | Insertion | Feed D ▷──────[+]──────● Delivery |
| | AddOp | Insert |
| | etc. | |

**Figure 5.11** The manual creation of assembly plan at an intermediate step

| BOM: | Actions Lib | Assembly Plan: |
|---|---|---|
| | Feeding | Feed C ▷──────┐ |
| | Delivery | │ |
| | Insertion | Feed D ▷──────[+]─────┐ |
| | AddOp | Insert │ |
| | etc. | Feed E ▷──[ ]──────[+]──────● Delivery |
| | | AddOp   Insert |

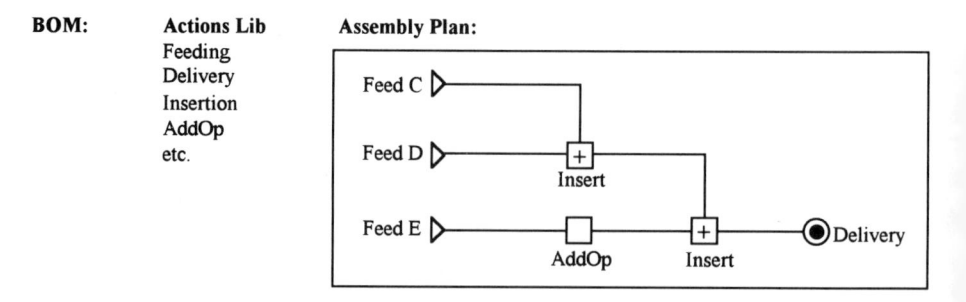

**Figure 5.12** The manual creation of assembly plan at the final step

## (iii) Edition of the assembly tree

Ordering actions is usually not sufficient. Some data are missing and, therefore, the assembly plan is not ready to be sent to the other modules. For this reason, other functionalities have been defined in order to completely define the assembly tree. These are:

- *Definition of a base part.* Every component of the BOM has to be either receiving components (base part) or placed on components (supplied part). A couple of components which are neither "base" nor "supplied" are called "base part undetermined" and the operator must to choose which is what. In other words, he has to remove some ambiguities on operation precedences (A on B or B on A?). The module automatically checks the base part indeterminacy.
- *Addition of reorientations actions.* We make here a commonly accepted hypothesis (see Delchambre, 1992) that all the assembly machines work with top-to-down movements. Therefore, two assembly actions using different assembly directions, e.g. X+ and X−, need to be separated by a reorientation action, which is automatically proposed to the user.
- *Addition of actions.* Some types of actions, not related to the assembly itself but nevertheless mandatory in an assembly line, can be added to the assembly tree: inspecting, painting, printing, or modifying the details of a process, such as adding a waiting time inside an adhesive bonding process.
- *Development of clusters of actions.* A cluster of components can be used in the assembly line by two means: either a special equipment realises the assembly operation simultaneously for all components of the cluster (e.g. screwing) or the cluster is (partially or totally) ordered by the operator in order to be assembled sequentially by different equipments. The operator can specify such ordering between components of a cluster.
- *Moving of actions.* For some reasons (mistakes, variants of a process, ...), the operator must have the possibility to modify the assembly tree by moving any of the operations already placed on the drawing.

## (iv) Storage of assembly trees

The operator has the possibility to define several assembly trees:

- either from scratch, with different ideas in mind on the look of the assembly plans; in this case, all plans are different and not related to each other.

- or from the same basic idea, but implicating different developments (e.g. some ambiguities remain); in this case, the plans are hierarchically or historically linked.

**(v) Enhancement of the manual procedure (semi-automatic scenario)**

This section presents which automatic functionalities are implemented in the editor and how:

- The geometrical precedence constraints—for each placement of action violating such a constraint, a warning is sent to the operator.
- The technological precedence constraints—for each placement of action violating such a constraint, a warning is sent to the operator.
- The editing of the reorientations—to check that two consecutive actions using a different assembly movement are separated by a reorientation action.

### 5.4.5 The Automatic Procedure

The goal of this section is to describe the automatic generation of assembly plans. The module takes as input the Bill of Material (BOM), the Bill Of Actions (BOA), the Bill of Precedence Constraints (BOC) and the performance indices. With these data, it computes a restricted set of assembly trees, sorts and displays them to the operator.

Then the operator compares the displayed assembly trees, chooses one of them and edits it. The chosen tree is then able to be sent to other modules. Some of these functionalities have already been defined, but others are still to be described; this will be done in the next sections:

- the computation of assembly trees;
- the performance indices;
- the visualisation and the choice of a specific assembly tree.

**(i) Summary of the method**

The automatic generation of assembly tree is a graph search, i.e. a loop where a new possible state is computed from a given state and where the computation begins again from this new state.

A state represents a *partial assembly tree*, i.e. a tree in which some parts are assembled and other are still separated. A whole assembly tree is a particular type of partial assembly tree, where all parts are assembled.

The computation of a state, arising from another one, is the aim of Section (ii) and uses the concept of "hyperarc". This concept has been introduced by Homem de Mello (see Homem de Mello and Sanderson, 1990, 1991).

This kind of problem (graph search) is combinatorial. In the assembly planner, the combinatorial explosion is reduced by several means:

- use of precedence constraints (both geometrical, technological and user-defined), already defined above;
- use of performance indices, described in Section (v).

If the number of partial assembly trees is not restricted, the system will be faced with some well-known features related to combinatorial problems: a huge number of solutions, exceeding both the memory of a computer and a reasonable computation time, and too numerous spurious solutions.

We define the *level* of an hyperarc as the number of components involved. So, the automatic method can be summarized in four steps:

1. compute the hyperarcs of a given level, ranging from 2 to the number of components in the final product;
2. compute the partial assembly trees of the same level;
3. select the best partial assembly trees of this level;
4. run the loop again, in the upper level.

The method produces as output a restricted set (from 10 to 30) of evaluated and sorted assembly trees.

**(ii) The hyperarc**

The computation of hyperarcs and the associated rules are inspired from Delchambre (1992).

A *hyperarc* is defined by a possible (dis)assembly between two subassemblies and is often written as "SA = SA1 + SA2". We define the *level* of a hyperarc as the number of components in SA. Beginning with the level number 2 (i.e. 2-parts subassemblies) and after that considering higher levels (i.e. 3-parts subassemblies, 4-parts, etc.), the process runs until the final assembly is obtained.

With the help of precedence constraints (coming from geometry, technology and the user), many of these decompositions will be discovered as invalid and, therefore, their number

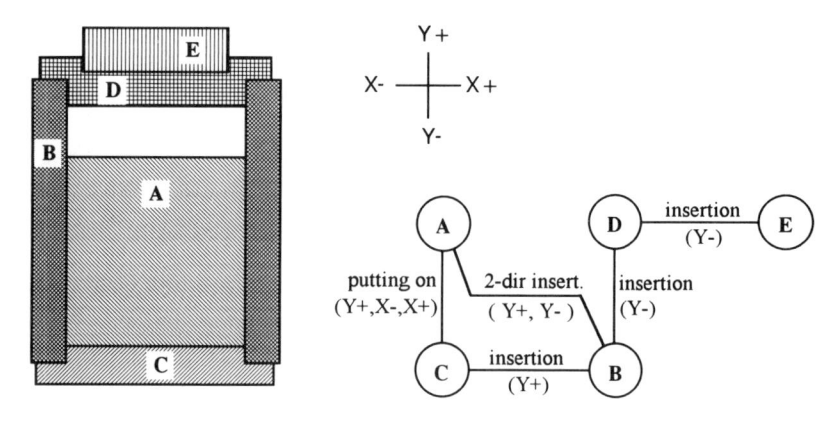

**Figure 5.13** A simple product and its relations

will decrease strongly. An example follows, based on Figure 5.13.

Considering all couples of components, the system generates the following hyperarcs. In the left-hand column are the subassemblies to be decomposed and in the right-hand column the decompositions obtained by dividing the original graph into two subgraphs (these must remain continuous).

| | | | |
|---|---|---|---|
| [a,b] | ⇒ [a] | + [b] | no. 1 |
| [a,c] | ⇒ [a] | + [c] | no. 2 |
| [b,c] | ⇒ [b] | + [c] | no. 3 |
| [b,d] | ⇒ [b] | + [d] | no. 4 |
| [d,e] | ⇒ [d] | + [e] | no. 5 |

Then, the system computes the following hyperarcs using three and four components:

| | | | |
|---|---|---|---|
| [a,b,d] | ⇒ [a] | + [b,d] | no. 6 |
| | [d] | + [a,b] | no. 7 |
| [a,b,c] | ⇒ [a] | + [b,c] | no. 10 |
| | [b] | + [a,c] | no. 11 |
| | [c] | + [a,b] | no. 12 |
| [b,d,e] | ⇒ [b] | + [d,e] | no. 13 |
| | [e] | + [b,d] | no. 14 |
| [a,b,c,d] | ⇒ [c] | + [a,b,d] | no. 15 |
| | [d] | + [a,b,c] | no. 16 |
| | [a,c] | + [b,d] | no. 17 |

#  Assembly Planning

$$
\begin{aligned}
[a,b,d,e] &\Rightarrow [a] + [b,d,e] && \text{no. 18}\\
&\phantom{\Rightarrow} [e] + [a,b,d] && \text{no. 19}\\
&\phantom{\Rightarrow} [a,b] + [d,e] && \text{no. 20}\\
[a,b,c,d,e] &\Rightarrow [c] + [a,b,d,e] && \text{no. 24}\\
&\phantom{\Rightarrow} [e] + [a,b,c,d] && \text{no. 25}\\
&\phantom{\Rightarrow} [a,c] + [b,d,e] && \text{no. 26}\\
&\phantom{\Rightarrow} [d,e] + [a,b,c] && \text{no. 27}
\end{aligned}
$$

With the help of the precedence constraints, we have removed:

$$
\begin{aligned}
[a,b,c,d] &\Rightarrow [a] + [b,c,d]\\
[a,b,c,d,e] &\Rightarrow [a] + [b,c,d,e]
\end{aligned}
$$

The computation of hyperarcs then stops because we obtain the whole product. It is obvious that, for a greater assembly, the computation has to go on.

Note that hyperarcs nos. 8, 9, 21, 22 and 23 have been removed because they involve subassemblies [b,c,d] and [b,c,d,e] from which it is impossible to build the final product. Addition of powerful rules restrict the number of hyperarcs further still (for more details, see Delchambre, 1992).

**(iii) The generation of partial assembly trees**

Obtaining a partial assembly tree is simple once hyperarcs have been generated. The system proceeds in merging the hyperarcs previously computed. Once all hyperarcs of a given level have been generated, then the generation of partial assembly trees of the same level starts. The last level does not produce actual partial assembly trees, but whole assembly trees, which have the final assembly as root and the single components as leaves.

Considering the previous example, the system generates this list of trees:

[a,b,c,d,e]=[c]+[a,b,d,e]
    [a,b,d,e]=[a]+[b,d,e]
        [b,d,e]=[b]+[d,e]
            [d,e]=[d]+[e]
⇒ tree no.1 = [24,18,13,5]

[a,b,d,e]=[c]+[b,d,e]
    [b,d,e]=[a]+[b,d,e]
        [b,d,e]=[e]+[b,d]
            [b,d]=[b]+[d]

⇒ tree no. 2 = [24,18,14,4]

[a,b,c,d,e]=[c]+[a,b,d,e]
    [a,b,d,e]=[e]+[a,b,d]
        [a,b,d]=[a]+[b,d]
            [b,d]=[b]+[d]
⇒ tree no.3 = [24,19,6,4]

[a,b,c,d,e]=[c]+[a,b,d,e]
    [a,b,d,e]=[e]+[a,b,d]
        [a,b,d]=[d]+[a,b]
            [a,b]=[a]+[b]
⇒ tree no.4 = [24,19,7,1]

and so on, we obtain:

tree no. 5 = [24,20,5,1]
tree no. 6 = [25,15,6,4]
tree no. 7 = [25,15,7,1]
tree no. 8 = [25,16,10,3]
tree no. 9 = [25,16,11,2]
tree no. 10 = [25,16,12,1]
tree no. 11 = [25,17,2,4]
tree no. 12 = [26,2,13,5]
tree no. 13 = [26,2,14,4]
tree no. 14 = [27,5,10,3]
tree no. 15 = [27,5,11,2]
tree no. 16 = [27,5,12,1]

These results are complete solutions, because the left-hand term is the final assembly and the right-hand term contains only separate components.

**(iv) The selection of partial assembly trees**

At a given level, after generating partial assembly plans, the system get firstly too many trees (several dozens) and furthermore too many unrealistic trees. Therefore, from this big set, a subset of 10 to 50 partial assembly trees have to be extracted in order to be considered at the next level (or displayed to the operator, if the system is at the last level).

The automatic method uses a multicriteria approach (outranking relations), which we can summarise by Figure 5.14, which displays two indices and sixteen partial assembly trees,

# Assembly Planning

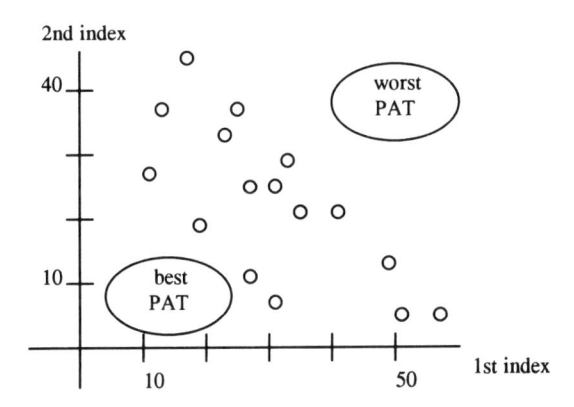

**Figure 5.14** An overview on the multicriteria analysis

called "PAT" in the following. Each vertex represents a PAT. Each axis represents the values of an index. We suppose that the smaller the value is, the better it is.

In order to perform this task, the system considers *performance indices*, which help to select the "good" hyperarcs and to eliminate the "bad" ones. With the help of performance indices, some of the partial assembly trees built in this way will *not* be taken into account. At each step, only the best ones (from the point of view of the performance indices) will be kept for further computations (Figure 5.15).

In this example, the criterion taken into account is the number of reorientations involved in the assembly process. The less there are, the better it is.

The next section describes the different performance indices implemented in the Assembly Planner. This means that the method has to take into account *simultaneously* several indices to select plans, which can be competitor or contradictory. The main question is: "Can I select a set of plans and, if so, which ones?" The method we have developed to perform this task is outside the scope of this book, but we can give general rules to the reader.

- If a plan is "very good" for at least one index, then the plan is selected. This implies that the range of this index is wide enough and thus this rule is applicable only at a high level.
- If a plan is "globally good" for all indices, then the plan is selected.

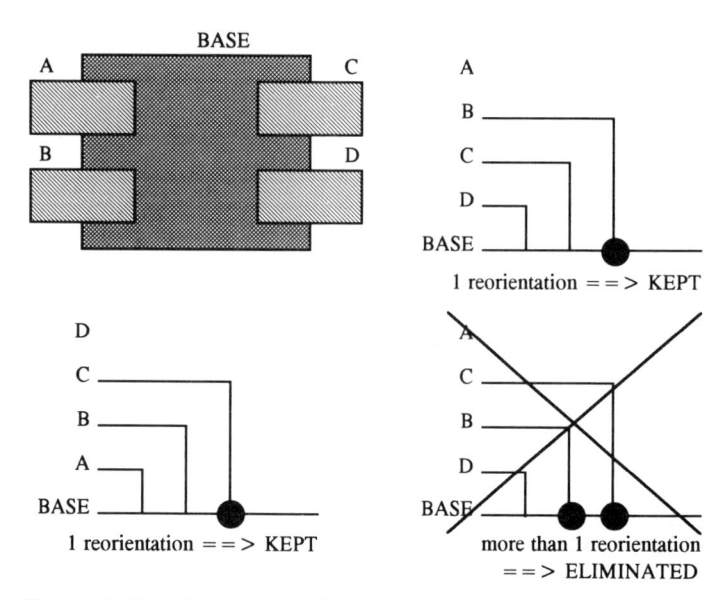

**Figure 5.15** An example of tree selection.

- If a subset of plans are equivalent from the point of view of all indices (i.e. represented by the same vertex), then we consider only one of them (and eliminate the other ones). A possible reason of this situation is that two plans can be different only because of a few components; for instance, two given plans are identical, except the component A is placed just before the component B on the first plan, and the opposite on the second one. Unfortunately, both A and B have no special feature, leading to the computation of different indices, and therefore the two plans are represented by the same vertex in the space of performance indices.
- If a plan is worse than another one for all indices, then that plan is eliminated (Figure 5.16).
- If a plan is "very bad" for an index and "not so good" for the other ones, then that plan is also eliminated (Figure 5.17).

After applying these rules, the remaining PATs are selected for the next level (Figure 5.18).

**(v) The performance indices**

The performance indices are used before, during and after computation of the assembly plans. Their aim is to reduce the com-

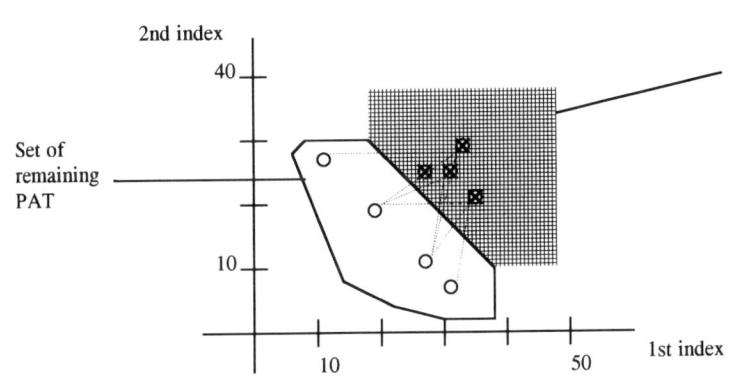

# Assembly Planning

**Figure 5.16** A manner to eliminate a PAT

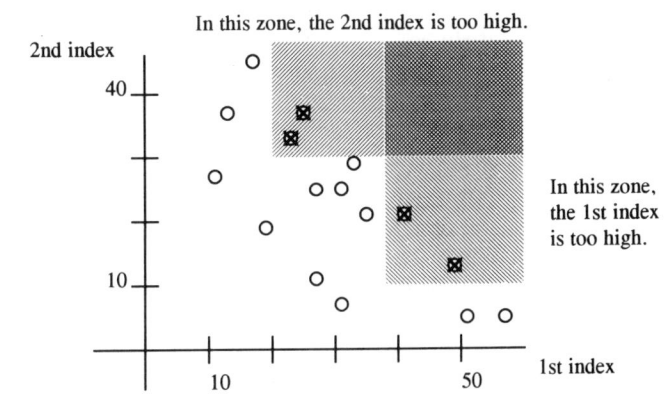

**Figure 5.17** Another manner to eliminate a PAT

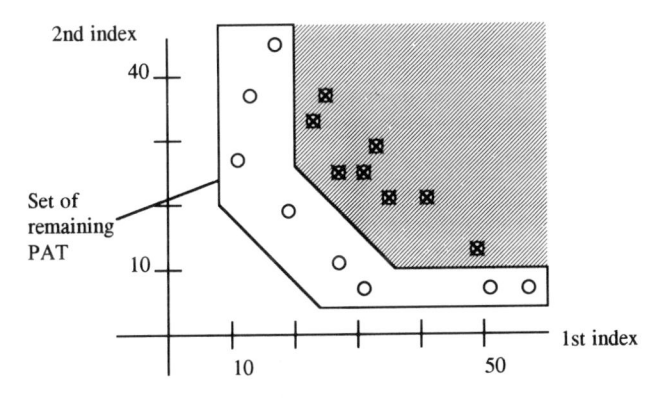

**Figure 5.18** The set of remaining PATs

binatorial explosion related to all graph search problems, as assembly planning. They are based on the assembly technology, taking the following features into account:

- the number of reorientations;
- the stability of intermediate subassemblies;
- the parallelism between operations;
- the earliest and latest components.

They are operator-selectable. Other indices could be defined, but they have to be specified by the user.

The *minimisation of the number of reorientations* is a index which leads to minimizing the number of times subassemblies are reoriented while the product is being assembled. It applies whether assembly is robotised, automatic or manual. When, for instance, a robot can do nothing but vertical insertions (SCARA), each time the assembly direction changes the fixture must be changed or the subassembly repositioned in the fixture. In the case of manual assembly, it is also more efficient to group parts according to their assembly direction. This approach reduces handling during the various operations.

For all these reasons, it is important to minimize the number of direction changes for component insertion during assembly.

The *maximisation of the stability of intermediate subassemblies* is an index for which the purpose is to avoid forming unstable subassemblies when building the product. In other words, it avoids mating parts which tend to disassemble spontaneously under the action of gravity or vibrations. The appearance of instabilities during assembly complicates, for example, the fixtures and robot grippers. Clearly, if a part tends to disassemble spontaneously in a fixture, it must be stabilised by external means (e.g. pneumatic jacks).

On the other hand, if two subassemblies are to be joined and one is unstable, transfer of the latter requires a sophisticated gripper to prevent the corresponding connections from coming undone. The stability criterion can provide useful information to the product design module. If a product cannot be assembled following this criterion, then this means that the product is perhaps ill-designed. Indeed, one of the basic concepts of good assembly design is to prefer insertion of parts which, once in place, are maintained in place by their physical contact with the other parts.

We have defined several types of (in)stability, but their

# CAD Method for Industrial Assembly

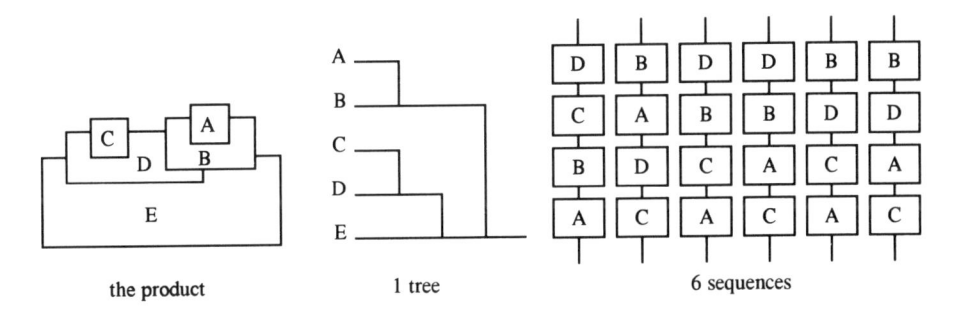

**Figure 5.19** A measure of the parallelism

description is outside the scope of this book (for more details, see Delchambre, 1992).

The *maximisation of the parallelism* is a third index. In an assembly tree, parallelism between operations is present: some subassemblies may be realised simultaneously and independently before being mated. The number of sequences in a plan is a certain measure of its parallelism, as shown in Figure 5.19.

In some cases, the user wants to obtain a plan representing as many assembly sequences as possible, so as to allow as many degrees of freedom as possible for scheduling according to the production means of the line.

Please note that the parallelism of a tree can be described by other means than the number of sequences:

- the width of the assembly tree;
- the depth of the assembly tree;
- the summation of the differences between the number of parts of each subassembly involved in each assembly operation.

The index related to *earliest and latest parts* is due to the fact that, as already explained above, it is often best, for instance, to place expensive components as late as possible. They are called "latest parts". The opposite idea is applicable, for instance, for stout components: it is advisable to assemble these early. They are called "earliest parts".

The idea of this criterion is to use such qualitative attributes to compare decompositions and partial assembly trees. An action assembling an earliest part has to be placed as soon as possible in the partial assembly tree. An action assembling a latest part has to be placed as late as possible in it.

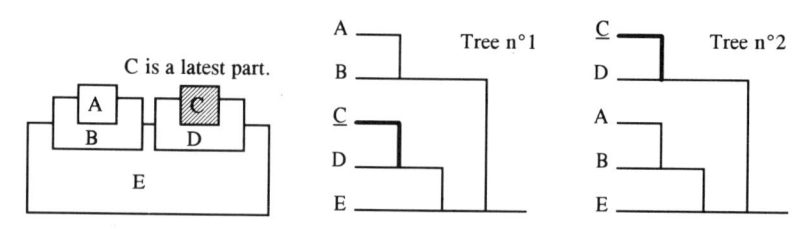

**Figure 5.20** An example of latest part

On this example, C having to be placed as late as possible, the tree no. 2 is better than the no. 1, because the operations involving C can be completed later. On no. 1, {A,B} is to be placed after {C,D}. On no. 2, {C,D} is to be placed after {A,B}: this is better.

**(vi) Visualisation and choice of assembly plans**

We consider that an automatic method is not so able as the operator to choose the "best" assembly tree. We think that the final choice has to come from him. The aim of this section is to describe how he can choose the "best" assembly plan.

Despite the graph search, the precedence constraints and the performance indices, the user may be faced, at the end of the computation, with several dozens of assembly plans. To allow an easy choice by the user, an automatic evaluation of these plans must be performed.

This evaluation (after computation) is strongly related to the performance indices (as for the computation itself). In the proposed method, we display to the operator an array containing the values of the following indices, for every remaining assembly plan:

- the number of product reorientations during assembly;
- the number of unstable sub-assemblies;
- the degree of parallelism of the assembly plan;
- the position of earliest and latest components.

This array of values is the basic data for permitting the operator to select the "best" assembly plan. The operator can choose, among this array, the assembly tree he wants to visualise. Figure 5.21 shows an example.

| assembly tree number | 251 | 124 | 26 | 3 | 247 | 841 | 120 | 15 |
|---|---|---|---|---|---|---|---|---|
| nb of reorientations | 2 | 3 | 2 | 3 | 3 | 2 | 3 | 3 |
| degree of stability | 6 | 6 | 5 | 5 | 6 | 6 | 4 | 4 |
| degree of parallelism | 8 | 10 | 12 | 11 | 9 | 10 | 12 | 10 |
| position earliest parts | 25 | 26 | 21 | 22 | 21 | 21 | 24 | 25 |
| position of latest parts | 15 | 12 | 15 | 14 | 12 | 12 | 14 | 14 |

**Figure 5.21** An example of a selection array containing performance indices

By this general view, the operator can easily visualise the differences between trees and therefore easily choose the assembly tree to keep for further purposes.

Each assembly tree will be displayed separately, as already explained in the manual procedure.

## 5.5 Conclusions

The state of the art shows that the proposed assembly planner represents a significant contribution in the field of assembly planning. Indeed, many features involved in this tool (specification of an editor, concurrent engineering, ...) are not often the aim of other assembly planners.

The product model is well defined, including all features necessary to assembly planning: qualitative and functional attributes, rule-based technology module, etc. We note that the use of a feature-based CAD system is required.

The concept of *actions* (i.e. the process model) is fully defined.

The macro-actions are the basic data of the assembly plan editor, which is also fully defined and is the basis of the manual creation of an assembly plan.

An automatic scenario is also fully specified, containing precedence constraints generation, assembly trees generation and selection, and performance indices.

Last but not least, the integration (as of the data structure) of the assembly planner with other modules as product design or resource planner has been one of the hardest tasks of this project, due to the complexity and the inter-relations between modules.

## 5.6 Bibliography

Bullinger H.-J. and Thaler K. (1989) Planning of flexible assembly based on graphs—an integrated approach, *10th Intl. Conf. Assembly Automation* (ICAA), pp. 343–350, Springer-Verlag, Berlin.

Bullinger, H. J. and Richter, M. (1991) Integrated design and assembly planning, *Computer-Integrated Manufacturing Systems*, 4(4), 239–247.

Esprit III Project No. 6562 (1992) Systematic Concurrent Design of Products, Equipments and Control Systems (SCOPES), Part 1: *General Project Overview*, June.

Hoffman, R. (1990) Assembly planning for B-rep objects, *Proc. 2nd Intl. Conf. Computer Integrated Manufacturing*, IEEE Robotics and Automation Society, pp. 314–321, New York, May.

*Méthodes et outils pour le montage et le démontage des roulements*, SKF—Information produit 300, pp 2–11.

Thaler, K. (1991) Computer-aided planning for assembly-enhancing productivity with computer-based support, *Advances in Production Management Systems*, pp. 343–349, Elsevier, North-Holland, IFIP, Amsterdam.

Wafflard, A. and Delchambre, A., Van Melsen, D., Peng Y. M., Huang Y. and Zhou, Z. (1991) *High Level Language for Robotized Assembly*, Research Report, EEC DG XII–China Cooperation, University of Brussels and Shanghai Jiao Tong University, 1988–1991.

Vincke, Ph. (1992) *Multicriteria Decison-Aid*, Wiley, Chichester.

## 5.7 References

Baldwin, D. F., Abell, T. E., Lui, M. C. M., De Fazio, T. L. and Whitney, D. E. (1991) An integrated computer aid for generating and evaluating assembly sequences for mechanical products, *IEEE Trans. Robotics and Automation*, 7(1), 78–94.

Barnett, J. A. (1990) A system engineering approach to automated assembly planning, *Proc. 2nd Intl. Conf. Computer Integrated Manufacturing*, IEEE Robotics and Automation Society, pp. 314–321, New York, May.

De Fazio, T. L., Abell, T. E., Amblard, G. P. and Whitney, D. E. (1990) Computer-aided assembly sequence editing and choice: editing criteria, bases, rules and techniques, *IEEE Intl. Conf. Robotics and Automation*, pp. 416–422.

Delchambre, A. (1990a) Knowledge-based control of a flexible assembly cell, ISATA, 23rd Intl. Symp. Automotive Technology and Automation, *Advanced Automotive Manufacturing*, Vol. II, stream B, pp. 384–391, Austria, December.

Delchambre, A. (1990b) Conception Assistée par Ordinateur de Gammes Opératoires d'Assemblage, PhD in Applied Sciences, May.

Delchambre, A. (1992) *Computer-Aided Assembly Planning*, Chapman & Hall, London.

Delchambre, A. and Gaspart, P. (1992) KBAP: an industrial prototype of knowledge-Based assembly planner, *IEEE Intl. Conf. Robotics and Automation*, Nice, France, May.

Delchambre, A. and Wafflard, A. (1990) A pragmatic approach to computer-aided assembly planning, *IEEE Intl. Conf. Robotics and Automation*, pp. 1600–1605, May.

Delchambre, A. and Wafflard, A. (1991) An automatic, systematic and user-friendly computer-aided planner for robotized assembly, *IEEE Intl. Conf. Robotics and Automation*, pp. 592–598, April.

Delchambre, A., Wafflard, A. and Gaspart, P. (1991) Knowledge-based 'assembly by disassembly' planning, *13th IMACS World Congress on Computation and Applied Mathematics*, Dublin, Ireland, July.

Dini, G. and Santochi, M. (1992) Automated sequencing and subassembly detection in assembly planning, *Ann. CIRP*, **41**(1), 1–4.

Eskicioglu, H. (1990) An expert-system approach to the assembly planning of roller chains, *Engineerial Applications of Artificial Intelligence;* **3**, 306–312.

Heemskerk, C. J. M., Reijers, L. N. and Kals, H. J. J. (1990) A concept for computer-aided process planning of flexible assembly, *Ann. CIRP*, **39**(1), 25–28.

Henrioud, J. M., Bonneville, F. and Bourjault, A. (1991) Evaluation and selection of assembly plans, in E. Eloranta (ed.), *Advances in Production Management Systems*, pp. 489–496, North-Holland, IFIP, Amsterdam.

Homem De Mello, L. S. and Sanderson, A. C. (1990) Evaluation and selection of assembly plans, *IEEE Intl. Conf. Robotics and Automation*, pp. 1588–1593, May.

Homem De Mello, L. S. and Desai, R. S. (1990) Assembly planning for large truss structures in space, *IEEE Intl. Conf. Robotics and Automation*, pp. 404–407, May.

Homem De Mello, L. S. and Sanderson, A. C. (1991a) A correct and complete algorithm for the generation of mechanical assembly sequences, *IEEE Trans. Robotics and Automation*, **7**(2), 228–240.

Homem De Mello, L. S. and Sanderson, A. C. (1991b) Two criteria for the selection of assembly plans: maximizing the flexibility of sequencing the assembly tasks and minimizing the assembly time through parallel execution of assembly tasks, *IEEE Trans. Robotics and Automation*, **7**(5), 626–633.

Mascle, Ch. (1990) *Approche Méthodologique de Détermination de Gammes par le Désassemblage*, Master Thesis, Département de Microtechnique, Ecole Polytechnique Fédérale de Lausanne, Switzerland.

Mascle, Ch. and Figour, J. (1990) Methodological approach of sequences determination using the disassembly method, *Proc. 2nd Intl. Conf. Computer Integrated Manufacturing*, IEEE Robotics and Automation Society, pp. 483–490, New York, May.

Mascle, Ch. (1992) Détermination automatique "à propos" des sous-assemblages, *Systèmes de production discontinue*, **26**(2), 167–192.

Miller, J. M. and Stockman, G. C. (1990) Precedence constraints and tasks: how many task orderings? *IEEE Intl. Conf. Robotics and Automation*, pp. 408–411.

Park, J. H., Kwon, D. G. and Chung, M. J. (1991) Framework for the evaluation and selection of assembly plans, *IEEE, Proc. Intl. Conf. Industrial Electronics, Control and Instrumentation* (IECON), pp. 1215–1221, Kobe, Japan, October.

Requicha, A. A. G. (1989) Representation of Tolerances in Solid Modelling: Issues and alternatives approaches, pp. 3–22.

Santochi, M. and Dini, G. (1992) Computer-aided planning of assembly operations: the selection of assembly sequences, *Robotics and Computer Integrated Manufacturing*, **9**(6), 439–446.

Wolter, J. D. (1990) A constraint-based approach to planning with subassemblies, *IEEE Intl. Conf. Robotics and Automation*, pp. 412–415, May.

Wolter, J. D. (1991) A combinatorial analysis of enumerative data structures for assembly planning, *IEEE Intl. Conf. Robotics and Automation*, pp. 611–618, April.

Zhang B. and Zhang, L. (1990) The automatic generation of mechanical assembly plans, *PRICAI*, pp. 668–672.

# 6 Resource Planning

## 6.1 Introduction

Once the marketing department has set the general framework of the future production of the company (what kind of product should be produced and in what quantities), the design department has designed the product (using, in particular, the tools of design for assembly) and the assembly planning has outlined the nature and succession of operations necessary to assemble the product, the resource planning carries over those decisions towards the design of facilities required for the final assembly of the product.

As mentioned in Section 2.4, the goal of resource planning is to decide *who* should assemble what, *where*, and using what *means*.

In other words, resource planning is concerned with the selection of *production means* adequate for performing all the assembly operations specified by the assembly planning (in a sequence compatible with the *assembly plan* it supplies) on parts specified by the design for assembly, while meeting the production volumes set by marketing. The position of the resource planning inside the design level of the architecture, proposed in Chapter 3, is depicted in Figure 6.1.

Note that a production workshop can be set up following various topologies—line, cells (islands), combination of several lines, pure jobshop (isolated workstations). However, in the context of *assembly*, the most usual topology is that of an *assembly line*, due to the fact that the output of the workshop consists of *one product* (with possible variants). This product takes shape gradually, starting with one part (usually called the base part), the remaining parts being attached at the various workstations visited by the product. Thus the topology of the line connecting the various workstations is a natural topological

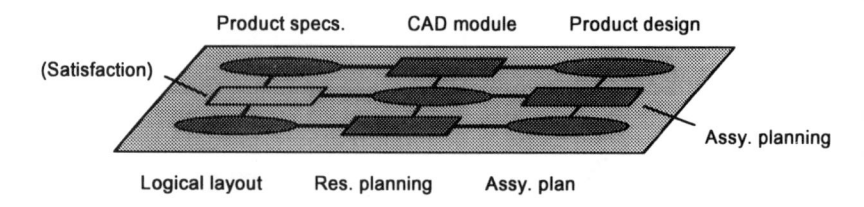

**Figure 6.1** Situation of the module within the CAD method for industrial assembly architecture

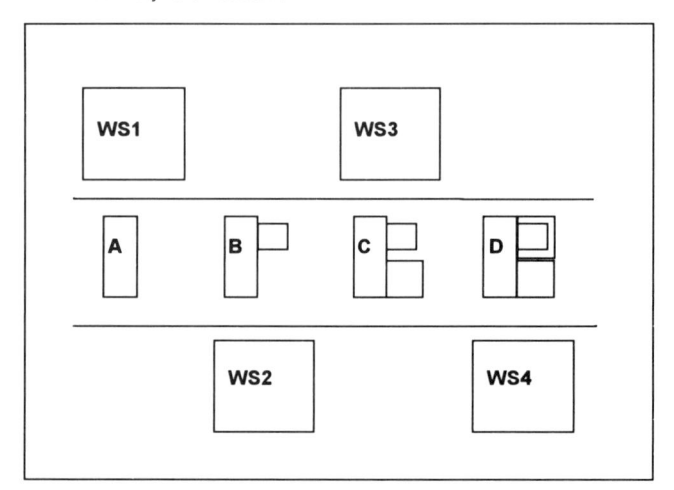

**Figure 6.2** The assembly line concept

expression of the assembly process. Consequently, the mission of the resource planner described in this chapter is the design of assembly lines.

The concept of assembly line is illustrated in Figure 6.2. Four workstations (WS1 through WS4) are connected by a conveyor. Each workstation adds one or more components (parts) to the product being assembled, which makes it undergo various phases of completion—the four phases A, B, C and D in the case of the example. The last workstation adds the last part and the finished product leaves the line, i.e. the line ends.

Of course, all the workstations along the line work in parallel, which means that at any given instant, there are as many products in the line (i.e. on the conveyor) as there are workstations.*

---

*Supposing there are no buffers in the line.

While the conveyor connecting the workstations usually moves continuously for mechanical reasons, it would be clearly impractical to attempt to attach parts to a moving product. Hence in front of each workstation there is an *indexing table*, a device that subtracts the product from the movement of the conveyor and fixes it in a precise position during the time necessary for performing the assembly operation(s) carried out at the workstation. Thus for the purposes of the resource planner, we can assume that the conveyor actually *stops* during that time.

Since all the workstations work in parallel, it follows that the pace of the whole assembly is given by the slowest workstation, i.e. the one that takes the longest lapse of time to perform its task(s). On the other hand, if there is a faster workstation in the line, then that workstation must wait for the slower one, otherwise there would be a continuous accumulation of products in front of the slow workstation.

Consequently, we can adopt the following abstract model for the assembly line. In the beginning, the conveyor is still, all workstations have a product in front of them and start their respective assembly operations. After a lapse of time called the *cycle time* $T_C$, the conveyor moves in the sense of the assembly (from the left to the right in Figure 6.2) at infinite speed, moving all products to the following workstation, after which it stops again. This makes up an assembly cycle, one finished product has been collected at the end of the line and a new one has been introduced at its beginning. A new cycle can start.

Clearly, all workstations must have finished their tasks at the end of the cycle, i.e. no workstation can take longer than $T_C$, the cycle time. On the other hand, since the conveyor is still during that time, any workstation faster than $T_C$ will be *idle* up to then.

The cycle time thus determines the pace at which the finished products are delivered by the assembly line. As such, it determines the production volume. However, it is the marketing policy of the company which dictates how many products should hit the market at any given period. Consequently, the cycle time is one of the constraints under which the line designer has to work: it is one of the input data for the resource planner.

The line depicted in Figure 6.2 is a classic *linear* one, meaning that there is a logical beginning and end to it: the assembly starts on the first workstation and, after visiting each of the workstations exactly once, the product is finished at the last one. A more general set-up, a *loop*, would allow the product to skip any of the workstations (thus ending up unfinished at the "last" one) and return with the conveyor to the "first" one any number of times.

## 6.2 Users' Needs

The ultimate objective of an industrial company is maximisation of profits. Thus the above activities are aimed at producing the best possible products (so they can be easily sold, for a good price) with the smallest possible investment necessary (so the margins are high). However, today's market conditions, with its fierce competition and consumer exigencies, has put an emphasis on two aspects of a production:

- The *time-to-market* delay has to be as short as possible, so that the company can quickly adapt to changing customer demands, thus making its products more appealing to the customer than those of the competition.
- At the same time, the production of the product must be as cheap as possible, so that it can be offered to the customer for a price that makes it competitive.

Resource planning addresses these two aims. First, it shortens the *time-to-market* delay by allowing the company to decide very quickly how to set up an assembly line, once the operations to be carried out there have been established. Second, the resource planner *optimises* the resulting assembly line, in order to make the cost of the production as small as possible.

As pointed out above, resource planning addresses the problem of deciding of "who should do what, where, and using what means", given an assembly plan and production rates. This encompasses several problems, which solution leads to a specification of an assembly line:

- decide which operation(s) of the assembly plan will be performed at which workstation along the assembly line, while complying with precedence constraints over the operations (specified by the assembly plan), and the constraint of cycle time. Thus the *composition* of the workstations is established.

    Note that the cycle time we mention here is the cycle time of the assembly line, derived from the production volume target. Note that the two figures are the same in the case when the whole production is performed on one assembly line, but the cycle time is a multiple of the production interval* when several lines are allocated for the production. On the other hand, some workstations on an assembly line can be multiplied (several running in parallel), thus effectively multiplying the cycle time for those workstations while maintaining the original cycle time for other workstations and the line as a whole.

- For example, suppose the marketing department has decided that 35 040 units of a product should hit the market over the next year. Should all those products be assembled on one line running eight hours a day, then the production volume for that line would be 35 040/365 = 96 products a day, i.e. 12 per hour, implying a cycle time of 60/12 = 5 minutes, equal to the production interval. Should two lines assemble the product in parallel, then the cycle time for each of them would be 10 minutes.
- Decide what production means, i.e. what *equipment*, should be used by each of the workstations. Needless to say, the equipment used by a workstation depends on the operation(s) performed there, i.e. on the composition of the workstation. Together, these two problems define the *logical layout* of the future assembly line, that is, they define the resources necessary to *modify* the product during the assembly process.
- Decide where each of the workstations should be placed on the shop floor, i.e. what should be the *physical layout* of the assembly line. The positions of the workstations on the shop floor influence the choice of the equipment necessary to *transport* the products from one station to the other, as well as the flow of tools, people and any other supplies (electricity, compressed air, etc.) throughout the shop.

## 6.3 State of the Art

The bibliography available on resource planning is rather scarce, which could be explained by the simultaneous difficulty and simplicity of the subject. Indeed, even the simplest variants of the underlying *optimisation* problem turn out to be NP-complete, i.e. computationally very difficult. On the other hand, very simple, nearly trivial heuristics can be found that often produce good solutions at least for simplified versions of the general problem, making it a "simple" one.

---

*The inverse of the production volume, i.e. the production time of one unit.

The simplest version of the problem is the well-known *bin-packing problem*, where the aim is to pack unidimensional "objects" of various sizes into as few identical "containers" of a given capacity as possible. This problem is relevant for resource planning in the following way: assuming that the durations of all operations to be performed on the line are fixed, they can be taken for the unique dimension of the "objects". Then casting the cycle time as the capacity of the bins, the problem becomes one of distributing the operations among workstations in such a way that no workstation overflows the cycle time and the minimum number of workstations is used. Thus the bin-packing problem leads to a minimisation of the number of workstations along the line. Note, however, that this simple formulation does not take into account the precedence constraints between operations, which amounts to assuming that a product can be assembled by any succession of the operations.

Garey and Johnson (1979) show the strong NP-completeness of the problem (thus casting a serious doubt on an exact and efficient algorithm), yet show that the simple first fit descending (FFD) heuristic uses not more (but sometimes not less) than 11/9 of the optimum number of bins.

Much more sophisticated approximation algorithms exist, e.g. the reduction approach of Martello and Toth (1990a) based on the notion of *dominance* (Martello and Toth (1990b) gives a survey of several others). A branch-and-bound method based on the dominance criterion, the MTP procedure of Martello and Toth (1990b), is considered by many to be the most powerful technique available for the bin-packing problem.

On the meta-heuristic side, (Falkenauer and Delchambre, 1992) developed a *Grouping Genetic Agorithm* (*GGA*), a Genetic Algorithm (Holland, 1975; Goldberg, 1989) heavily modified to suit the structure of grouping problems, yielding far better results than the FFD heuristic.

Finally, the recent *hybrid grouping genetic algorithm* of Falkenauer (1995), born from a marriage between the dominance criterion of Martello and Toth and the GGA of Falkenauer and Delchambre (1992) constitutes a method yet more powerful than the MTP procedure.

Fitted with acyclic *precedence constraints* (Sacerdoti, 1977), the bin-packing problem becomes the *line-balancing problem*.

A modification of the FFD heuristic (augmenting the sizes of the objects with the sizes of all predecessors) yields a simple heuristic recommended by Chow (1990). Given its nature, its performance is similar to the FFD heuristic.

Gutjahr and Nemhauser (1964) model the problem as a weighted directed network of *feasible subsets* (subsets of tasks that can be performed on one workstation), where the shortest path through the network corresponds to the optimal solution of the line-balancing problem. While the problem of finding the shortest path is a relatively easy one (e.g. Sedgewick, 1984), the number of nodes in the network grows exponentially with the size of the instance, making it unusable for large problems.

Peng (1991) attempts to take advantage of the additionnal constraints to obtain an exact solution. However, the inherently combinatorial nature of the problem seems fatal even to his sophisticated dynamic programming techniques, especially when the precedence constraints are loose.

Falkenauer and Delchambre (1992) generalize their bin-packing GGA to obtain a fast algorithm supplying high-quality approximate solutions of the line-balancing problem. One advantage is its ability to handle problems with sparse, even empty precedence constraints, thanks to its bin-packing "ancestor". Another advantage lies with the fact that the Genetic Algorithm is a *gradual improvement* method, thus giving the user the possibility, when the computational resources (time in particular) allow it, to extend the search and continue to improve the currently available solution.

To conclude the review of the approaches to the line-balancing problem, let us note that the mechanism of complying with precedence constraints proposed by Falkenauer and Delchambre (1992) applies equally well to the Hybrid GGA of Falkenauer (1995), leading to an equally powerful algorithm for line-balancing.

A further generalization of the problem's definition relaxes the fixed-length tasks, replacing them with tasks whose duration depends on the equipment or number of workers assigned to each workstation. This yields *line-balancing with resource dependent task times*.

Building on the exact algorithm of Gutjahr and Nemhauser (1964) for the line-balancing problem, Faaland *et al.* (1992) use heuristics for building only those nodes of the network which have a reasonable chance to lead to an optimal solution, thus yielding a reasonably fast approximation algorithm.

Finally, the problem can be generalized by relaxing the fixed set of tasks and precedence constraints, replacing them with a set of *variants* (of a base product), all of which have to be assembled on the same assembly line.

Holmes (1987) addresses this problem, using a method similar to the one of Gutjahr and Nemhauser (1964), i.e. finding the shortest path in the network of feasible workstations. However, given the implicit enumeration of all possible workstations, her method is impractical for large problem instances.

## 6.4 Functionalities and Methodology

### 6.4.1 Introduction

As pointed out in Section 6.2, there are three major tasks in resource planning: elaboration of the logical layout of the line, which consists in a simultaneous *selection of equipment* used for each operation and *distribution of operations* among workstations along the line, and the subsequent elaboration of the physical layout of the line, i.e. deciding about the disposition of the workstations, conveyor(s), possible buffers, etc. on the shop floor.

The two steps of logical layout design are handled by two submodules of the resource planner, one handling the selection of equipment (discussed in Section 6.4.2 below) and the other optimising the distribution of operations among workstations of the assembly line (Section 6.4.3). The topic of physical layout design is discussed in Section 6.4.4.

### 6.4.2 Equipment Selection

**(i) Introduction**

It is a well-known fact that a substantial part of the effort spent in designing an assembly line consists in searching for data pertaining to equipment available for the operations in the assembly plan. Indeed, in order to choose adequate equipment, designers have to scan numerous catalogues, contact many producers of equipment, keep themselves informed of the latest developments in traditional as well as new technologies, etc., without ever being sure that all data worthy of consideration have been collected. One novelty of the resource planner presented here is that it aims at eliminating most of this "administrative" burden, by making available all (or at least most) of the useful information in form of an *equipment database*.

## (ii) Equipment modelling

A database is only useful if there exists a well-defined procedure allowing retrieval of only the data *pertinent* in a given context. In the case of an equipment database, this means that a practical procedure must be defined for matching a given assembly *operation* with all the *equipment* that could carry out that operation, possibly subject to some additional criteria. In short, the ultimate aim here is a tool that would "answer" the question "What are the equipment adequate for the operation X?" with a list of equipment retrieved from the database.

The problem of equipment selection can be thus formulated as follows.

1. Given an assembly operation, what can be said about it that would be *relevant* for the choice of *equipment* accomplishing the operation?
2. Given an equipment, what can be said about it that would be *relevant* for characterisation of the *operations* the equipment can be applied to?
3. In order to lead to an effective *matching procedure*, the answers in points 1 and 2 must match each other.

Our solution to the problem is to model the equipment and the operations using attributes that define the relevant characteristics in points 1 and 2 above.

In trying to define the appropriate attributes, one can observe that the description of an assembly operation can be given with various degrees of generality. To illustrate this point, let us suppose the product contains a riveting joint operation among two parts A and B, and let us see the various characterisations of that joint:

Level 1   Join A and B
Level 2   Rivet A and B
Level 3   Rivet A and B using the rivet of type R
Level 4   Rivet A and B using the rivet of type R, performing a "one shot" technology
Level 5   Rivet A and B using the rivet of type R, performing a "one shot" technology, preheating the rivet to 180° C.
⋮
Level $\Omega$   Rivet A and B ... using the machine M

As can be seen, going from one level of the description of the *joint* to a higher-numbered one involves more knowledge to be dealt with, as the operation is more fully specified. On the other hand, each level restricts the choice of the *equipment* that could possibly be used to accomplish that joint.

Note also that the different levels of description of the joint involves knowledge ranging from purely *product design*-related down to the *production method*-related, with basically all intermediate mixes possible. Indeed, the designer's ultimate goal is simply to join the two parts together in such a way that the joint *satisfies the constraints* applied to it when the product is used. There is no doubt that if the designer could do with the Level 1 only, he or she would gladly do so. On the contrary, the ultimate choice of the equipment which will be mounted on the shop floor is a matter usually handled by the production department which has little knowledge (or influence) necessary to design or modify the product.

The lower the level number of description (i.e. conceptually higher the description) of the joint operation available to the resource planner, the broader is the choice of the equipment which could *a priori* be used to make that joint. As an extreme, the Level 1 (the lowest-numbered, conceptually the highest) alone would imply that *any* equipment could be used to join the two parts A and B together. Of course, this is without taking into account information that could perhaps be extracted *automatically* from the design (i.e. CAD drawing and specs) of the product. For example, it seems clear that if A and B are plastic parts, arc welding could not be used to join them. However, such an automatic extraction of knowledge would most probably involve enormous development and maintenance costs on the part of the software (e.g. an expert system). As an example, in the case of the above joint, many things influence the choice of the rivet to use, such as the efforts applied to the joint given the overall geometry of the product, the function of the joint (should A be allowed to turn with respect to B or not, should the joint conduct electricity, . . .), the environment which the product will be used in (corrosion. . .) and so on.

Most of this information would be very difficult to extract automatically from the product alone. However, most of the information is directly available to the product designer and can be thus simply input by him or her if the appropriate questions are raised by the system during the product design phase.

On the other hand, some of the information leading to the choice of the particular machine M is outside the scope of the product designer's knowledge and is the matter of the *line* designer. Such information cannot, of course, be supplied by the *product* designer.

We thus proceed in two stages which roughly match the division between *design* and *production* as it exists in today's industry. This warrants a practical usability of this part of the resource planner.

In a first stage, the *product design stage*, the *product* designer of the product will answer certain questions specifying, to some degree, the assembly operation itself. This translates into attributes that characterise the assembly technology used in that operation. These attributes are matched by those of the equipment that can perform the operation.

In a second stage, the *equipment selection stage*, the *line* designer selects, among all the equipment that are relevant with respect to the assembly operation, one or more equipment that meet his or her needs in the context of the particular assembly line being designed. These needs are expressed as additional attributes an equipment must possess.

In conclusion, the key to a successful (i.e. *practical*) resource planner lies with the appropriate choice of the *level of detail* of the information which should be supplied during the design phase of the product on the one hand and, possibly, during the design phase of the line on the other hand.

We have first tried to produce a uniform "questionnaire" applicable to any kind of assembly operation. Our aim was to define a *mask* which would serve as a sieve which would only let "pass" the appropriate equipment. The advantage of such a unified approach would be the simplicity of the resulting *uniform structure* of the data involved. That, however, proved unrealistic, as the assembly operations are simply *too diverse*. In other words, different criteria apply to different kinds of operations and equipment. For example, it makes no sense to specify "one or two heads" (an important information for riveting, for instance) when arc welding is involved. In addition, there turned out to be assembly technologies that are more "delicate" than others, requiring more information for a reasonably restrictive specification of the process/equipment.

Consequently, our idea is to proceed following a *tree* of questions, each time giving a *finite choice* of possible answers, in order to ensure that the answers will lie in the scope of the system and will be properly handled. This will also ensure a reasonable ease of use of the system, as selecting from a short list of possible answers is easier than "inventing" an answer.

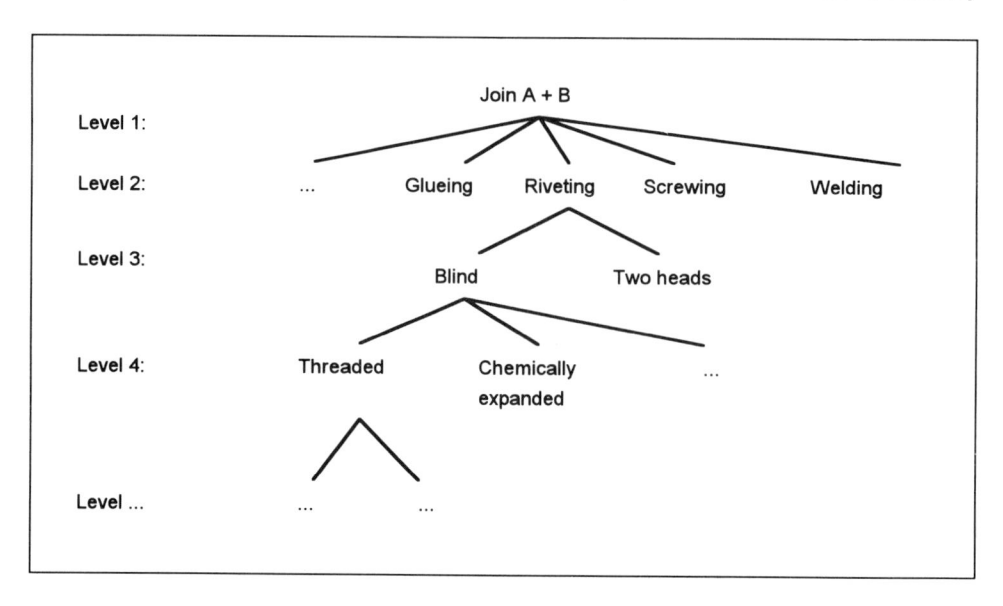

```
Level 1:                    Join A + B
                         ╱    ╱  ╲    ╲
Level 2:        ...   Glueing Riveting Screwing  Welding
                              ╱   ╲
Level 3:                   Blind   Two heads
                         ╱   ╲   ╲
Level 4:           Threaded Chemically  ...
                            expanded
                   ╱  ╲
Level ...        ...   ...
```

**Figure 6.3** Example of attribute tree characterising an assembly operation

At each point of the question tree, the answer given will lead to the corresponding subtree of questions. This will ensure that the questions asked will be coherent with the answers already given. To illustrate the approach, a (slightly more realistic) part of the tree relevant to the above riveting example above is given in Figure 6.3.

As pointed out above, the level of detail of the information supplied by the product designer should be chosen in such a way that as much as possible of the information useful for equipment selection will be obtained, yet such that no information will be requested that would be too *process* related and out of reach of the *product* designer. It is certain that at least a complete specification pertaining to the *product* (i.e. a complete specification of the assembly *technology*) should be supplied by the product designer, e.g. the precise description of the rivet to be used in the above example.

**(iii) Equipment retrieval**

From the information processing point of view, once the operations and the equipment have been suitably modelled using attributes they possess, extracting the relevant equipment from an equipment database is an easy task. Indeed, one only has to follow the successive answers in the question tree to retrieve the set of desirable attributes. These can be then matched against the attributes of the equipment in the database.

The process of retrieval of equipment relevant to a given operation for a given assembly line is iterative. Starting with the attributes specified by the product designer (i.e. those identifying the *technology* used in the operation), the information available at that point is used to extract all the equipment that match it, by a scan of the equipment database. The resulting list of equipment is presented to the line designer for verification. If the list turns out to be too long, or contains equipment not suitable in the context of the line being designed, the line designer uses his or her knowledge to specify the desired properties of the equipment more fully, thus constraining the possible choice of the equipment. This yields an enlarged list of attributes the equipment must possess, and the equipment database is scanned again with the enlarged list of criteria. This process of gradual "zeroing in" on the appropriate equipment is repeated as long as the line designer deems necessary.

Note that the second step of the equipment selection, i.e. the check of the list of available equipment, leaves the final word to the line designer. The extent of restrictions imposed on the equipment can thus range from none (*any* equipment relevant for the technology specified by the product designer is eligible) to a complete selection of one particular equipment.

Such a "Man in the loop" approach is necessary. Indeed, there is always a danger that some error has been introduced, be it in the data in the equipment database or in the answers given to the questions in the question tree, and in any case, it is extremely difficult to model every single aspect of the problem that could be relevant under any circumstances. The final check by the line designer will filter out any such errors.

**(iv) Cost and duration of an operation**

The main conclusion concerning the characterisation of an assembly process is that a uniform representation of the relevant attributes is impossible—a "piecemeal" approach, considering separate classes of assembly processes (technologies) is necessary. A similar conclusion can be reached with respect to determination of operation *duration* and *cost*. For instance, it is obvious that the duration of a screwing operation will be computed differently from a stapling operation, since there is no such parameter as the length of the thread in the case of stapling.

One way of handling this problem is to consider *approximations* of the real functions giving the cost and duration. Indeed, industrial experience shows that the major factors having an influence on these values usually boil down to a few variables in simple (usually linear) formulae. For instance, the duration of an electrode welding operation is closely approximated by $l/s$, where $l$ is the length of the weld and $s$ the speed of advance of the electrode.

## 6.4.3 Logical Layout Optimization

**(i) Problem definition**

Once the equipment retrieval task is accomplished, i.e. once each *operation* was assigned one or more *equipment* and for each couple [operation,equipment] the corresponding cost and duration of the operation was computed, the second module of the resource planner can be called upon in order to find the *most economic* logical layout, i.e. the distribution of the operations onto workstations along the line and the equipment assigned to each of the operations.

A valid logical layout must possess the following properties:

- All the operations specified by the assembly plan must be carried out (i.e. be assigned to a workstation and be allocated to adequate equipment).
- The precedence constraints specified by the assembly plan must be complied with, i.e. there must exist at least one order of the workstations in the resulting line under which the following holds: for each couple of operations O1 *precedes* O2 assigned onto workstations S1 and S2 respectively, either S1 *precedes* S2 in the line, or S1=S2 (the operations are performed on the same workstation).
- The constraint of cycle time must be complied with, i.e. for each workstation in the line, given the equipment assigned to each of its operations, the sum of the durations of all the operations must not exceed the cycle time. Note however, that the cycle time considered here is the local one, relative to a given workstation. Indeed, as we pointed out above, work-

stations can be *multiplied*, in which case the *local* time constraint is a multiple of the *global* cycle time.

Note that the above are necessary properties of a *valid* and *complete* logical layout. The resource planner *outputs* only valid ones, but that does not mean that in the process of elaboration of a logical layout, the constraints cannot be *temporarily* transgressed. Of course, the resource planner possesses the necessary means to validate and, if necessary, correct, a logical layout.

Three ways of elaborating a logical layout are available: a manual procedure, a fully automatic procedure, and a mixed one, i.e. a manual procedure made easier by automating some aspects of the work.

The automatic procedure must be preceded by the equipment selection described in Section 6.4.2 above. It is optional for the manual and mixed procedures.

## (ii) The manual procedure

**The overall procedure** The manual procedure is the "poor man's resource planner": it enables the end-user to create a logical layout of the future assembly line without recourse to the optimisation algorithms described in Section (iii) below or even without consultation of an equipment database described in Section 6.4.2 above.

The overall procedure is one of *gradual improvement* of a logical layout. This is achieved with the help of a suitable graphical representation that will guide the line designer towards the important aspects of the line he or she is working on.

The strategy of gradual improvement has one important aspect: it enables the line designer to *review* and possibly *improve* an arbitrary logical layout, should it be a default one, one proposed by a colleague, or one supplied by the Automatic Procedure described in (iii). In short, the manual procedure constitutes the necessary guarantee that the *man stays in charge*.

**The graphical representation of a logical layout** A good man–machine interface is the *sine qua non* of a successful manual procedure. It must represent the design problem being solved in such a way that the designer has a quick *overview* and *access* to all its important aspects.

In the resource planning task, the important features of a solution (i.e. a logical layout) are

- the total cost of the line;
- the costs of equipment assigned to individual operations;
- the time spent on each workstation, as compared to the cycle time.

Note that the last important characteristic of a logical layout, namely its compliance with the precedence constraints, can be displayed implicitly simply by insuring that *no* layout violating them will be handled. Indeed, these constraints are supplied by the assembly planner and as such constitute an external given of the problem, outside the realm of the resource planner.

In this context, it is useful to have a representation of the assembly plan handy, possibly on the same screen as the logical layout or, if this proves impossible due to the space constraint of the screen, in a separate window, so that the designer is able to access it with a single click of the mouse and/or make it overlap a currently uninteresting part of the logical layout window. Note however that the issue of representing assembly plans is a problem treated within the assembly planner module (see Section 5.4.4), so we will not come back to this topic here.

*Checking the precedence constraints*
*(i) The workstation precedence graph* In many stages of the logical layout design, the compliance of a solution with the precedence constraints has to be checked. The check consists in verifying that the workstations can be disposed along the future assembly line in such a way that for any operation $O$ in the assembly plan, the following holds: each of the predecessors of $O$ (if any) is performed either on the same workstation as $O$, or on a workstation that precedes the one $O$ is performed on.

Note that there can be more than one order of the workstations that satisfies the criterion, because the order on the set of operations need not be a *total* one. Hence what we want to know is this: given an assignment of operations to workstations, does there exist *at least one* order of workstations satisfying the precedence constraints?

Needless to say, we have to find an *efficient* algorithm for checking the criterion.

It turns out that this seemingly difficult problem (implicitly checking all possible orders of workstations, looking for the first feasible one) actually has an elegant and fast solution.

The solution becomes apparent when we realise that a violation of the precedence constraints leads to a *cycle* among the workstations. Indeed, if the constraints are violated for a given order of workstations along the line, the product being assembled has to move against the sense of the line conveyor, thus visiting at least one workstation several times. If this condition arises for *all* orders of workstations, then there is a cycle in the precedence graph among the workstations and, conversely, if there is a cycle in the workstation precedence graph then no order of workstations can satisfy all precedence constraints among operations.

Thus the criterion translates into the following algorithm:

1. Given a distribution of operations onto workstations and a set of precedence constraints among the operations, construct the workstation precedence graph (WPG) as follows:
   (a) Each workstation corresponds to one vertex in the WPG.
   (b) Given two distinct operations $O_1$ and $O_2$ on workstations $W_1$ and $W_2$ respectively, then if and only if $W_1 \neq W_2$ and there is a precedence constraint $O_1 \to O_2$, then there is an edge $W_1 \to W_2$ in WPG.

   Intuitively (and more simply), the WPG is the graph of the assembly plan, where all operations on a workstation have been "collapsed" into one node.
2. Detect the presence of a cycle in the WPG. If there is one, then the precedence constraints will be violated for any order of workstations along the line. Conversely, if the graph is acyclic, there exists at least one order of workstations that satisfies all precedence constraints.

*(ii) Computing ranks of a graph* Having constructed the workstation precedence graph (WPG) from the assembly plan and the assignment of operations to workstations, all we need to do to check for compliance with the precedence constraints is to detect an eventual cycle in the WPG. This is easily (i.e. efficiently) achieved by computing the ranks of the vertices of the WPG.

The rank of a vertex is defined to be the length (in terms of number of vertices) of the longest path entering the vertex. Thus a vertex that has no edge entering it has rank 0 (zero), a vertex entered by vertices issued from vertice(s) of rank 0 has itself rank 1, etc. Of course, a vertex that is part of a *cycle* has degree $\infty$ (infinity).

This immediately yields the following recursive procedure for computing the ranks of vertices in a graph G:

1. Set *current_graph* to G and set *current_rank* to 0.
2. Scan all vertices in *current_graph* and set the rank of all vertices with no incoming edges to *current_rank*. If no such vertex can be found (all vertices have at least one incoming edge or the *current_graph* is empty), exit.
3. Reduce the *current_graph* by eliminating all vertices of rank *current_rank* and all edges adjacent to them. Thus the new *current_graph* has fewer vertices and edges.
4. Set *current_rank* to *current_rank* − 1 and go to 2.

At the end of the procedure, the following holds:

- All vertices that have been eliminated in Stage 3 have their rank well defined.
- If there are any vertices left, then their rank is undefined (or infinite, as you prefer) and G (the original graph) *contains a cycle*, otherwise G is *acyclic*.

Needless to say, by taking the WPG for G in the procedure, this last point translates directly into our original criterion of compliance with precedence constraints.

It is easy to show that the above graph rank procedure runs in time of the order of $V^2$, where $V$ is the number of vertices in the graph G. Hence the compliance with the precedence constraints can be checked in time quadratic in the number of workstations, that is, in time bounded above by the *square of the number of operations*. That is very fast indeed—sufficiently fast to be performed *on-line*, as is necessary in an *interactive* environment.

*Displaying things* The graphical representation that seems to represent best the important aspects of a logical layout is the following (Figure 6.4).

- One position per workstation, the workstations spanning from the left to the right of the screen in the order of their position along the line.
- Each workstation is represented as a "heap" of boxes (rectangles), one box per operation, as in Figure 6.4. Its number is displayed below the heap.
- Each operation is represented as a box on the "workstation heap". Its *height* is proportional to the *duration* of the operation. Thus the workstation has the height proportional to the

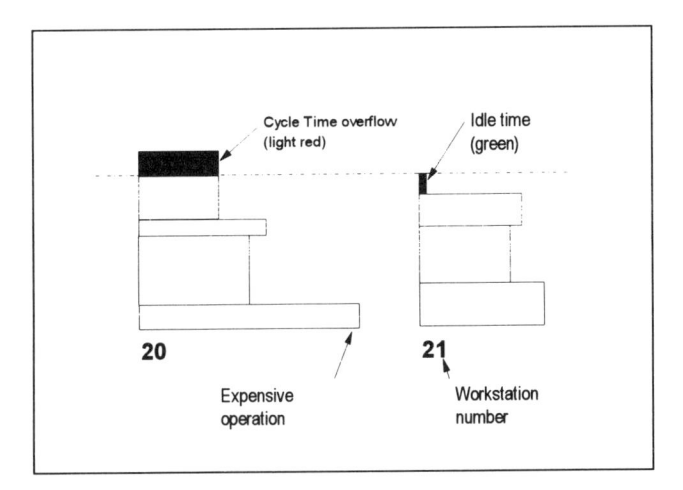

**Figure 6.4** Displaying workstations

total duration of the operations performed at that workstation. The *width* of an operation box is proportional to the *cost* of the equipment assigned to the operation.
- An eventual idle time on a workstation (underflow of the cycle time) is represented by a box of a special colour, e.g. green, put on top of the operation boxes.
- An eventual overflow of the cycle time is represented by drawing the boxes above the cycle time "waterline" in a special colour, e.g. light red.

Note that this representation indeed displays the important aspects of the logical layout:

- The extent to which the allowable time is exploited on each workstation is directly visible by comparing the heights of the workstations, with possible overflows signalled by the warning colour and unnecessary idle times signalled by large "idle time operations".
- High-cost operations are directly identified by their width.

*Accessing things* The "boxed" graphical representation of the logical layout allows for a quick access to its important aspects in view of a more detailed (textual) consultation of the data or a modification of the layout:

- Clicking the mouse on the number of a workstation opens a window giving the list of all the operations performed there, with their names and durations and costs of their respective equipment. The numbers of the immediate predecessor and successor workstations (according to the precedence constraints) are also displayed (see the last section on this page).
- Clicking the mouse on an operation opens a window displaying the name of the operation, the current equipment assigned to the operation with their costs (individual and total) and the duration of the operation. The numbers of the immediate predecessor and successor operations (according to the precedence constraints) are also displayed.

**Improving the layout**

*Selecting a new equipment* There are the following three possibilities for the choice of an equipment for an operation:

1. The equipment database is not used. The name of the equipment, its cost, and the resulting duration of the operation are input by the line designer.
2. Accessing the equipment database, the line designer selects *one* equipment according to the procedure described in Section 6.4.2 above.
3. Accessing the equipment database, the line designer selects *several* possible types of equipment according to the procedure described in Section 6.4.2 above. In that case, the system chooses the *cheapest* equipment that does not lead to an overflow of the cycle time. Note that the list of all the possibilities accepted by the designer is kept stored with the operation, in view of a future reoptimization of the workstation or the line (partial or entire).

*Moving an operation among workstations* Moving an operation from one workstation to another involves the following steps:

1. The designer selects an operation to move.
2. All the workstations that can accept the operation are determined, given the precedence constraints and the current composition of the workstations. This is easily done by supposing the operation already is moved to another station and detecting an eventual cycle in the resulting WPG (see pp 144–146). All workstations that *can* accept the operation are

clearly marked on the screen, for instance by highlighting their graphical representation.
3. The line designer selects the target workstation. This should be done with a mouse click (fast and intuitive) on the workstation to move to. However, the designer should still have the possibility to "consult" the contents of any workstation (see p 147, "Accessing things"), so that he or she can make a qualified choice of the target. This is easily achieved by two different ways of clicking, for instance a single click for a consultation and a double click to specify the target. Of course, should the designer attempt to select an invalid target, a warning is issued and the move is not allowed.
4. The operation is taken off the source workstation and put onto the target. This involves an update of the graphical representation of the two workstations.
5. According to the move having been performed, the order of workstations currently displayed may no longer comply with the precedence constraints. Indeed, the move has modified the assignment of operations to workstations, i.e. the workstation precedence graph, meaning that the set of all orders of workstations that comply with the constraints may have changed while still staying non-empty*. If that is the case, the workstations are reordered on the screen, after a suitable warning to the line designer (so they are not surprised by what is happening on the screen).

*Changing workstation position* As we have pointed out on p 144 (Section (i)), there can be (and usually are) several ways of disposing the workstations along the assembly line, in compliance with the precedence constraints. The line designer could thus be willing to alter a proposed order of the workstations, for instance in anticipation of the physical layout of the future line.

The interactive procedure involves the following.

1. Click on the number of the workstation to move, i.e. click where to move *from*. If the contents (i.e. the list of operations) of the workstation are not currently on the screen, it is displayed in order to enable the designer to verify that the workstation is indeed the one he or she wants to move.

---

*The non-emptiness of the set of possible workstation orders is guaranteed by only allowing moves towards workstations determined in step 2.

2. The permitted range of the move, i.e. the position(s) along the line the workstation can be moved to, is displayed, by two vertical lines delimiting the range. The way of computing this range is described below.
3. Click on the number of the workstation which is to be replaced by the one being moved, i.e. click where to move *to*. If the move is not a valid one (i.e. inside the permitted range), a warning is issued. Otherwise,
4. If the workstation is being moved towards the beginning of the line (to the left of the screen), all the stations between (and including) the *to* workstation and the *from* one are moved one position to the right (i.e. towards the end of the line) and the *from* workstation replaces the *to* one. Conversely, if the workstation is being moved towards the end of the line (to the right of the screen), all the workstations between the *from* one and (including) the *to* one are moved one position to the left and the *from* is placed where the *to* has been.

Note that moving *all* the workstations between the *from* and *to* positions does not modify their mutual order, i.e. the precedence constraints are not violated among them. The workstation being moved will not violate them either if the move is within the prescribed limits. These are very simply defined as follows:

- The *immediate predecessor* of a workstation W is the last (rightmost) workstation that contains an operation which is a predecessor of an operation on W. W must be placed *after* (on the right of) its immediate predecessor.
- The *immediate successor* of a workstation W is the first (leftmost) workstation that contains an operation which is a successor of an operation on W. W must be placed *before* (on the left of) its immediate successor.

**(iii) The automatic procedure**

**A two-tier approach** In order to justify our choice of the resource planner cost function, let us first discuss the difficulties encountered in its definition.

The cost function to be optimised by the resource planner's optimisation algorithm has for the ultimate objective the minimisation of the total *cost* of the assembly line. That cost is equal to the cost of all *equipments* used in the line. However, there are two classes of equipment that seem to require different treatment in the optimisation, namely (1) the equipment mounted on workstations and (2) the workstation (WS) itself. Indeed, the latter brings in the cost of the indexing table, the cost of the factory space, maintenance, etc. The difficulty of simultaneous minimisation of the two components of the total cost stems from the following.

If the number of workstations is fixed, the least expensive line is the one that uses the total time available as well as possible. That is, what has to be minimised here is the total time the workstations stay idle. Supposing that the speed of an equipment grows with its price, the tendency will be to select as *cheap* (i.e. *slow*) equipment as possible.

On the other hand, when the number of workstations is to be minimised, the objective is to "pack" as many operations on each workstation as possible. Thus the tendency will be to select as *fast* (i.e. *expensive*) equipment as possible.

The way out of this dilemma would be to define a measure of *quality* of a given workstation and promote, through a suitable cost function, solutions featuring good-quality stations. In the case of the pure line-balancing problem, a quality of a WS is simply defined as the extent to which the WS utilises the available *time*, since it is the only resource taken into account there. This approach is sensible: a well-performing WS maximises the time available on the *other* WSs, leading eventually to a global optimization of the number of workstations.

In the general resource planning problem tackled here however, a measure of quality is much harder to define. Since we want to minimise the total cost of the line, it is obvious that the measure of quality should be of the form performance/price. Indeed, we want to have the workstations as *cheap* as possible (so the total cost of the line is minimised), yet a workstation which does nothing is cheap but useless.

While the price is easily defined as the price of all the equipment attached to the WS, the performance measure is difficult to define. Intuitively, a well-performing WS is one that "does a lot of work", but what does that mean? Surely it cannot be the simple *number* of operations performed there, as there are more or less difficult operations. It cannot be the productive time (contrary of idle time) on the station, since two workstations can well use the same time while clearly having different performances, as shown in Figure 6.5. In short, an "absolute" measure of quality of a workstation is very difficult to define.

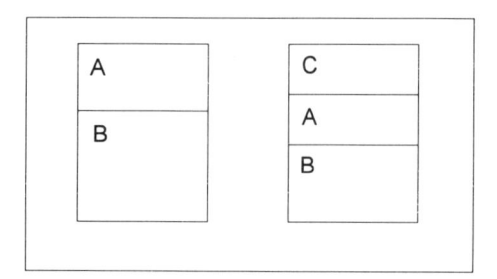

**Figure 6.5** Performance of a workstation

In order to remedy the above problems, and also to insure that *real costs* will indeed be optimised by the procedure, we have finally settled to a two-tier approach. We proceed in an *iterative relaxation* way as follows:

1(a). Select for each operation the *fastest* (most expensive) equipment possible. Run an algorithm for line-balancing, minimising the number of workstations required. The resulting number of workstations is the smallest possible for the line.
1(b). Select for each operation the *slowest* (cheapest) equipment and run again the line-balancing algorithm. The resulting number of workstations is the maximum useful for the line. Starting with the number of workstations obtained in 1(a),
2. Relax the current solution by adding one workstation. Run an algorithm optimising the cost of the line, given the current number of workstations.
3. Repeat point 2. until a satisfactory solution is obtained or the maximum number of workstations (obtained in 1(a)) is reached.
4. Among the solutions obtained for the various line "lengths", select the one that minimises the total cost of the line.

Note that for the first two subproblems (1(a,b)), the minimisation of the number of workstations given *one* choice of equipment per operation, the cost function of Falkenauer and Delchambre (1992) (see Section 6.3 above) can be used without modification. On the other hand, in the optimisation in point 2, a *constant* number of workstations is considered and the *real*

Resource Planning

cost of each workstation evaluated, the final cost function being the sum of the costs of all workstations in the line.

**Assigning equipment to one workstation** The two-tier approach does solve the problems discussed above, but introduces a new difficulty: once a set of operations has been assigned to a workstation, the cost of the workstation has to be computed. This problem is not trivial, because there can be several pieces of equipment available for each of the several operations assigned to the workstation, each choice of equipment implying a different cost and duration of the operation.

The problem is illustrated in Figure 6.6: a choice of more expensive (but faster) equipment for one operation can well free enough time for another operation to be assigned cheaper equipment, in such a way that the total cost of equipment attached to the workstation is lower.

Clearly, this is a combinatorial problem: in principle, *any* assignment of equipment to the operations on the workstation could be interesting, and there are exponentially many of them. Fortunately, there is the constraint of cycle time, i.e. the equipment must be fast enough to perform all the operations in that timeframe. This allows us to define a branch-and-bound procedure that efficiently explores the huge search space by following a tree depicted in Figure 6.7.

At any moment, a number of nodes of the tree are "open",

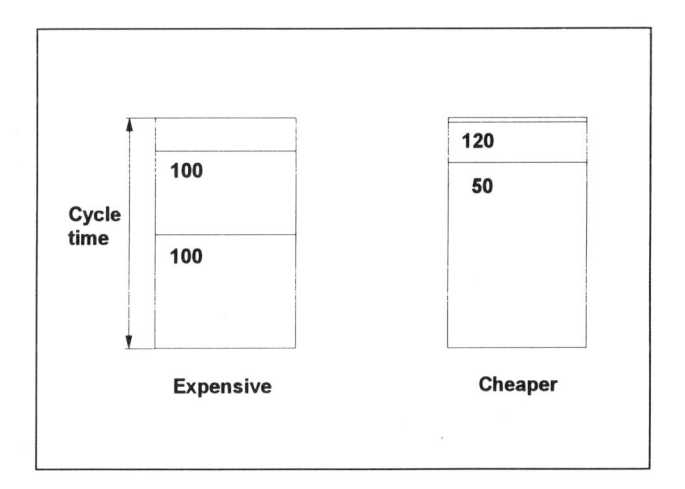

**Figure 6.6** Assigning equipment to one workstation

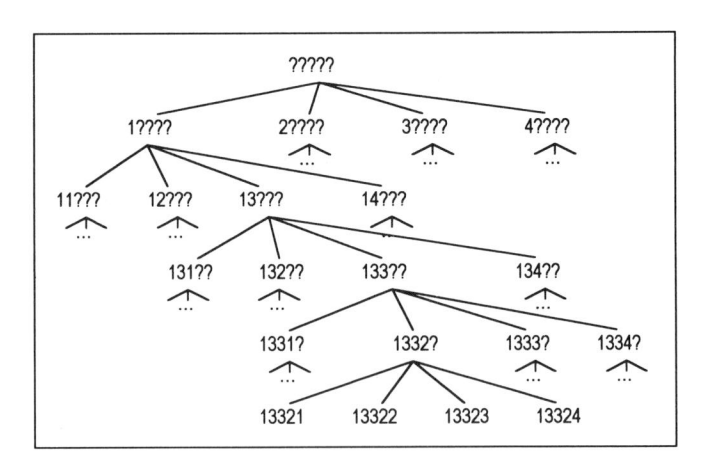

**Figure 6.7** The solution tree for equipment assignment

specifying a certain subset of the possible assignments of equipment to the operations. In the figure, there are five operations on the workstation, and there are four types of equipment available for each of them. Each node of the tree is "developed" to a certain degree, some of the equipment being fixed (a number in a node in the figure), others being so far undefined (a "?" in the figure). The algorithm proceeds iteratively, each time developing (i.e. fixing some choice of equipment to an operation) one of the nodes in the tree.

Of course, should *all* nodes of the tree be developed, the entire search would be explicitly explored, which would take extremely long execution times, making the algorithm unusable. Fortunately, it is possible to formulate efficient lower and upper *bounds* on the cost of solutions in any of the sets represented by the nodes of the tree. This allows us to cut the branches of the tree as soon as it becomes clear that no solution in the subtree issued from a node could be better (i.e. cheaper) than the best solution found so far. The search space is thus explored *implicitly* for most of it, making the algorithm quite efficient.

Without going into details of computing the bounds, let us say that they explore the constraint of cycle time. On one hand, *any* assignment of equipment must be fast enough, i.e. it must use equipment *expensive* enough to meet the "deadline", and this constitutes a lower bound on the total price of *any* equipment assignment. On the other hand, *some* assignment can fully exploit the available time, and the price of the cheapest equipment constitutes an upper bound on the price of *at least one* equipment assignment.

Of course, the bounds are just estimates, otherwise the problem would be trivial. However, as the nodes of the tree are being developed, these estimates become more and more precise, up to the point where they constitute a proof that *no* solution in the tree below a node can be useful (i.e. cheaper than the currently best solution). At that moment the branch stemming from that node is cut, and no further nodes in that branch are explored, constituting big economies in computation time.

**The overall algorithm** According to the point 2 in the two-tier approach described on p 150, an algorithm is necessary to find the best distribution of operations to a fixed number of workstations, with the aim of minimising the *cost* of the *whole* assembly line. Without going here into details of the algorithm, let us say that a grouping genetic agorithm (similar to the one for the line-balancing problem (Falkenauer, 1995)) is used to distribute the operations in compliance with the precedence constraints. The merit of each solution generated is evaluated by computing the best assignment of equipment to each workstation using the procedure outlined in the previous section, the cost of the whole line being the sum of costs of all workstations (i.e. of all equipment attached to the workstations in the line). One advantage of the genetic agorithm approach is that no "gradient" function is necessary: the cost function alone is capable of leading the genetic algorithm towards the optimum.

**(iv) The mixed procedure**

Beside the manual and fully automatic procedures described in Sections (ii) and (iii) respectively, the resource planner offers a *mixed* procedure to the line designer. This is the manual procedure where the designer solves parts of the design problem with the help of optimising algorithms.

Automatic procedures can be called for the following aspects of the logical layout elaboration:

- check of precedence constraints;
- optimal selection of equipment for a workstation performing operations assigned by the designer (i.e. the designer decides which operations to perform on the workstation, while the equipment that will be used is selected automatically);

- the same as 2, but with *some* of the equipment imposed by the designer;
- optimisation of a *partial* logical layout, i.e. the designer assigns *some* operations onto workstations, possibly imposing equipment for *some* of the operations, the remaining operations and/or equipment being optimised by the algorithm outlined in the previous section.

The mixed procedure probably constitutes the most user-friendly way of elaborating a logical layout: wherever the line designer wants to impose decisions, he or she does so using the user-friendly interface described in Section 6.4.3(ii). However, all parts of the design process that are tedious and where the designer does not have an *a priori* decision to impose, are carried out by the resource planner's algorithms. This way, the designer can concentrate on creative aspects of the design process. In addition, he or she can "play" with various aspects of the layout, assessing within short response time the impact of decisions. This leads to high-quality logical layouts, adapted to whatever special requirements the line must satisfy.

### 6.4.4 Physical Layout Design

Curiously, the physical layout design, i.e. the problem of deciding exactly *where* each of the workstations (i.e. the machines, the feeding equipment, etc.) should stand on the shop floor, is usually *not* considered to be a difficult problem by designers of assembly lines.

This result is surprising because from the computational point of view, physical layout *is* a hard problem. It has two aspects:

1. decide where to put each of the machines in order to allow *all* of them to stand on the allowable surface of the shop floor without "overlapping";
2. among the solution(s) of the previous problem, i.e. those arrangements of the machines on the floor that let all the necessary equipment inside the allowable space, choose the one that minimises the overall cost of the various traffics in the shop.

The problem is indeed a hard one: already its first aspect is NP-hard in the strong sense (Garey and Johnson, 1979), let alone the two together! It is thus rather curious to observe that the assembly line designer does not consider the problem, and thus an eventual tool for solving it, of a high priority in resource planning.

The little emphasis given to the physical layout problem by today's line designer can be attributed to two factors:

1. As far as the first aspect of the problem is concerned, the space allocated for an assembly line (typically the factory hall where the assembly will take place) is in most cases *sufficiently large*, so that the problem of placing all the machines and equipment inside that space is nearly trivially solvable "by hand".
2. Concerning the second aspect of the problem, the investment into the transportation system (conveyors) is regarded as only a small part of the overall investment. In particular, it is considered to be a secondary problem in comparison to the choice of equipments on the workstations. In cases where this is not the case (e.g. where AGVs are used as means of transportation), solutions obtained "by hand" following simple rules of common sense, are deemed to be of sufficient quality, i.e. such that an additional effort of optimization of the Layout would produce solutions which cost would be only marginally better than the one already at hand.

On the other hand, an automatic treatment of the physical layout design problem faces an additional difficulty besides the computational complexity of the optimisation: a quantitative assessment of the *quality* of a given physical layout, i.e. the definition of a suitable objective function to optimise.

This difficulty stems from the fundamentally *multi-criterion* nature of the problem. Indeed, it is extremely difficult to compare the cost of, say, plumbing for compressed air in comparison to the nuisance of having noisy equipments in close proximity to a meeting room. Of course, it is qualitatively clear that the latter should be placed as far from one another as possible, but whenever this implies long pipes for compressed air, the respective importance of the two criteria is extremely difficult to quantify.

All the above observations point to a manual editor as the tool of choice for elaboration of the physical layout of a workshop. Consequently, the CAD method for industrial assembly proposes a physical layout editor to complete a logical layout.

## 6.5 Conclusions

The resource planner described in this chapter constitutes a user-friendly tool for elaboration of high-quality logical layouts of assembly lines, i.e. it assigns operations to workstations along the line and selects equipment to carry out those operations, by minimising the *cost* of the line.

Note that even when working with incomplete data, i.e. in the context of concurrent engineering, the resource planner can provide useful information to the product/production means design team: an absence of equipment for a certain operation, or prohibitive price of any equipment capable to carrying out the operation, constitute a useful hint that the product should probably be redesigned using different assembly technologies. Such an advice can be usefully handled in particular by design for assembly. An impossibility to usefully combine together operations on workstations, leading to long idle times, constitutes a hint to the assembly planner that a different assembly sequence should be adopted.

Once the logical layout is elaborated, the physical layout editor is used to decide of the position of the workstations and their equipments on the shop floor. The transportation devices (conveyors, AGVs, etc.) and possible buffers are designed as well. At that stage, a complete model of the future assembly line has been obtained, and the behavior of the line can be *simulated* in order to assess the validity of the design.

## 6.6 References

Chow We-Min (1990) *Assembly Line Design, Methodology and Applications*, Marcel Dekker Inc., New York.

Faaland, B. H., Klastorin, T. D., Schmitt, T. G. and Shtub, A. (1992) Assembly line balancing with resource dependent task times, *Decision Sciences*, **23**, 343–364.

Falkenauer, E. (1995) A hybrid grouping genetic algorithm for bin packing, submitted to *Intl J. Computers and Operations Research*.

Falkenauer, E. and Delchambre, A. (1992) A genetic algorithm for bin packing and line balancing, *Proc. 1992 IEEE Intl Conf. Robotics and Automation*, May 10–15, Nice, France, IEEE Computer Society Press, pp. 1186–1192.

Garey, M. R. and Johnson, D. S. (1979) *Computers and Intractability—A Guide to the Theory of NP-completeness*, Freeman, San Francisco, CA.

Goldberg, D. E. (1989) *Genetic Algorithms in Search, Optimization and Machine Learning*, Addison-Wesley, Reading, MA.

Gutjahr, A. and Nemhauser, G. (1964) An algorithm for the line balancing problem, *Management Science*, **11**(2), 308–315.

Holland, J. H. (1975) *Adaptation in Natural and Artificial Systems*, University of Michigan Press, Ann Arbor, MI.

Holmes, C. A. (1987) *Equipment Selection and Task Assignment for Multi-*

*product Assembly System Design*, Master of Science in Operations Research Thesis, Massachusetts Institute of Technology, MI.

Martello, S. and Toth, P. (1990a) Lower bounds and reduction procedures for the bin packing problem, *Discrete Applied Mathematics*, **22**, 59–70, North-Holland, Elsevier, Amsterdam.

Martello, S. and Toth, P. (1990b) *Knapsak Problems*, Wiley, Chichester, pp. 221–245.

Peng, Y. (1991) *The Algorithms for the Assembly Line Balancing Problem*, Internal Report, CRIF Industrial Automation, August.

Sacerdoti, E. D. (1977) *A Structure for Plans and Behavior*, Stanford Research Institute, Elsevier, North-Holland, Amsterdam.

Sedgewick, R. (1984) *Algorithms*, Addison-Wesley, Reading, MA.

# 7 The Simulation Module

## 7.1 Introduction

Simulation has become an essential tool when dealing with manufacturing and assembly system problems, at the design, implementation and operation stages. During the design phase simulation assists the decision making process to define features, kinds of equipment, layout, control, scheduling strategies and contingency plans when failures occur. Furthermore, simulation is a useful way to test the actual control software of the planned system before implementing the real process (Garnusset *et al*, 1990).

Developments in simulation software such as animated graphics and the ability to model systems in detail have done much to make simulation methods accepted and increase the user confidence in the simulation results (Heger and Richter, 1994). Especially, the improved communication derived from a simulation model in graphic mode helps understanding, and the benefit is a better cooperation between different disciplines and departments.

Today the vast majority of simulation applications have centred on the off-line analysis of new or existing manufacturing or assembly scenarios. The aim was to examine the dynamic behaviour of a new plant or workshop when introducing new cells or different control philosophies and operation rules.

In an off-line implementation of a typical simulation model, operators are required to enter data (which could be the layout of the system, operation times, machine breakdown etc.) via keyboard and mouse. This may be acceptable for some applications; however, a better approach is to supply the simulation module with information about the layout already generated in the resource planner and furthermore with flow control rules and decisions as derived from the flow control module.

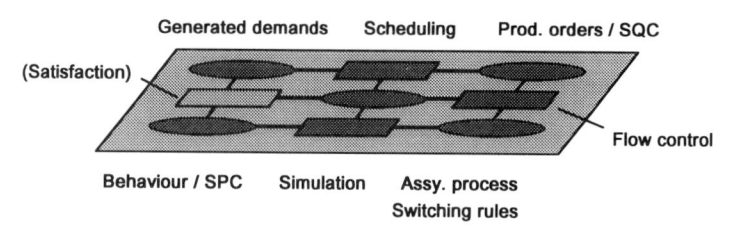

**Figure 7.1** Situation of the module within the CAD method for industrial assembly architecture

The simulation module represents the link between the off- and on-line modules of our CAD method for industrial assembly. The integration of the simulation module in the architecture of the CAD method is shown on Figure 7.1.

The simulation module accesses the layout database (not shown in Figure 7.1) and receives information about the assembly process by receiving the bills of work for each resource. It generates and changes dynamically the system state during a simulation run by creating events for the single elements and reacting on those events. In order to receive information about the traffic and routing of parts in the system, the simulation module accesses the local switching control rule of each switch as generated by the flow control module. Information about the cycle-time and parts to be produced are received via bills of work from the flow control module. The simulation module exchanges events with the flow control module. In case the local flow control information is not sufficient to make a routing decision, it will receive a global switching rule. The simulation module will react to events placed by the flow control module in the simulation event queue (such as polling of a resource).

The concurrent engineering approach attempts to increase the overlap between the activities of product design with the planning and development of the production process. *The role of simulation, as the validating and integrating factor to bridge the knowledge gap between the planning and production departments, is crucial in any concurrent engineering implementation* (Heger, 1994).

Simulation in a CAD method for Industrial assembly supplies the user with a simulated system in order to validate concurrently the solutions provided by the off- and on-line modules before implementing the real assembly system.

Using this simulation module, the user receives the following information about the dynamic system behaviour at a very early planning stage:

- proof of viability of the production system;
- potential bottlenecks in the assembly system;
- statements for possible scheduling and flow-control strategies.

The decisions of the off-line module are influenced by the simulator. Furthermore, it supplies the flow-control module with a simulated system in order to enable the user to complete the flow rules. The user is able to test the local and global flow rules before they are transmitted to the real layout.

## 7.2 Users' Needs

In order to develop the simulation module, industrial requirements were gathered by carrying out a simulation questionnaire and by bilateral interviews with the potential users.

They have a large range of different assembly scenarios:

- manual assembly in a workshop organisation;
- manual assembly, assembly line organization;
- mixed manual and automatic assembly;
- automatic assembly.

The industrial users have already gained experience using commercial simulation tools in order to validate the architecture of the workshop. Simulation experience was partly gained by verifying that the work load along parts of the assembly line conformed to the cycle time. Other simulation projects already carried out by the users were to study buffer sizes or to analyse the global production rate.

Regarding the users, the main drawback of commercial tools is their complexity and the problems that most of them have to be used by simulation experts and not by the engineers doing day-to-day work. Therefore the main requirement on the simulator from the industrial users was that the simulation tool should not require modelisation skills (Datin and Bovone, 1993). The users do not want to master the model behind the equipment in the workshop.

The build-up phase of the simulation model is the prime interest of potential users, since it is likely that the users are involved in the planning of the physical line. This requires a powerful interface from the CAD to the user to indicate the results of the search for logical layout. Hence, the interface from the user to the CAD has also to be powerful, in order to complete the results and to get a complete physical architecture.

The simulation tool is used in different stages of the design process.

- *At the beginning of product and process development.* Here the input of the simulation module will be a logical layout. The users wish to validate as early as possible an assembly scenario and refine this scenario until a complete definition is obtained. Information about the given cycle time and transport times between stations is very useful at this stage. The first model of the workshop should not require any modelisation effort from the user, since the logical layout itself is a result of the resource planning. However, this first simulation will allow the verification that the production rate is respected, which is the crucial parameter of any line behaviour. Very interesting features of a simulation tool at this stage are to detect possible bottlenecks in the equipment. This leads to a possible feedback to the resource planning, or in terms of operations, even lead back to the assembly planning. Here, the user requirement can be specified so that simulation speeds up the phase of delivering a first rough specification of the line and strengthens the proposal of a manufacturing solution by justifying quantitatively the appropriateness of the solution to the customer requirements.

  It is likely that the dynamic functions of the management of the manufacturing—scheduling and flow control—be already part of this simulation. For instance, in case an operation is performed by three stations in parallel and in derivation of the main flow, the rules that control the switches of the derivation are necessary for a simulation run. Therefore, the logical layout model may already be enriched through interfaces in order to define those functions.
- *Product and process development* At this stage the logical layout is enriched and completed. The transfer system will be described at a level that allows, for instance, a conveyor to validate the length of the conveyor and the number of pallets.
- *Production means building and setting up.* The interest of using simulation at this stage is to compare the results obtained in

the design phase with those obtained in run time, in order to analyse the differences and tune the functions.

Some relevant features of the simulation module, regarding the users point of view, will be described as follows:

- *Graphical representation.* The representation of the elements should be as simple as possible. Simple blocks representing the stations and arrows representing the flow are best suited to the description and user analysis. The production state should be available at any time. More information about a station should be obtained upon request.
- *Model description.* Three types of stations might be part of the workshop: manual, automatic and semi-automatic stations. For each of them, it is absolutely necessary to have the choice to associate a law that modifies the cycle time around a mean value. In case of automatic equipment, associate another law that simulates different kinds of problems: jamming, missing of parts in the feeding devices, breakdowns. Set-up times due to changes in the production must be integrated as well as maintenance times.

  As transfer between stations and between cell's conveyors, automated guided vehicles (AGVs) or manual transport might be used.

  Buffers are essentially studied in terms of capacity. Feeding problems can occur in assembly; therefore a corresponding law should be applied.

  Since for some users 80% of the production costs are generated by manual production, it is essential to be able to simulate manual production. From an organisational and economical point of view, to choose the right number of workers with sufficient qualifications is an important task in assembly systems. Simulation can help to specify their work, taking their schedules, breaks and qualification into account. Therefore it is necessary to include the modelling of workers in the simulation module. The model of a worker includes for instance his or her qualifications and work-time.
- *Simulation.* During the time of the set-up of the workshop model, a step by step simulation is necessary. A set of colours corresponding to the different states of the machines is more user-friendly than counter display. Since a graphical animation is an important time consumer, users require the simulator to have animation only as a request function.

## 7.3 State of the Art

The modelling and simulation of technical systems, especially production systems has increased a lot during the recent years and gained much attention (Chan and Smith, 1992; Eversheim *et al.*, 1994; Toncich, 1992). Accordingly the number of simulation systems available on the market and being developed has increased as well.

Most of the commercially available simulation systems are used to:

- prove the viability of a production system;
- find essential points in bottlenecks in the production system;
- predict the production process;
- derive statements for possible production strategies.

Those simulators are stand-alone solutions, which have to be supplied manually by data. In an off-line implementation of a typical simulation model, operators are required to enter data via keyboard and mouse. Those stand-alone solutions are different from our approach, where the simulator is one module in a concurrent design and development scenario and has a very extensive data exchange with the other modules. To our knowledge an integrated simulation tool, as being developed in this book, is unique.

Amongst the large number of papers published on simulation research during the last years some of them are describing problems that are related to our field of work. Spur *et al.* (1993) discuss possible techniques using simulation for product development and factory planning.

A procedure for layout design and material handling system selection for a synchronous manufacturing cell is described by Antonio and Egbelu (1991). The facilities layout and the material handling system are determined through the use of mathematical optimisation and discrete event simulation. However, a very specific solution is described, which can not be generalised into other types of assemblies.

The feasibility analysis of implementing a just-in-time (JIT) system in a manufacturing environment is an interesting approach for simulation (Chengalvarayan and Parker, 1991). The objectives of the implementation are to increase the product line utilisation and to reduce the work in progress (WIP) inventory. The machine breakdown time is high and the production schedule issued is very dynamic; the schedule is modified every two days. The study is done in the semi-automated industry, producing a composite product of three major elements. In order to carry out the study, computerised production schedules and production line data had to be collected. The production floor was converted into a simulation model, the simulation runs were represented in trace reports to verify and validate the program. The problem tackled in this paper is related to the scenarios described in this book. It seems an available integrated simulator system would have drastically reduced the time spent on data-collection and model building.

Wiendahl and Scholtissek (1993) describe a simulator to simulate workshop oriented manufacturing. Their simulator has the possibility to take over action plans and capacity information from other sources. A scheduling module is integrated in the system. The main advantage of this simulator is the possibility to take various scheduling strategies into account and to have a prospective test of the strategies before implementing them in the real system. The use of the simulator for the development and test of a scheduling module for electronic assembly is shown. There have been numerous contributions on the use of Petri nets (PNs) to design and build simulation tools. However, the traditional approach of PNs in manufacturing lies in the development of control models for automated manufacturing systems (AMSs). More recent research on PNs in AMSs is reviewed by D'Souza and Khator (1994). The application of PNs to the real-time control of AMS is still a present challenge to researchers. Several papers (Boucher *et al.*, 1990; Teng and Black, 1990; Hatono *et al.*, 1991) describe the recent developments in this field.

Okada and Ushio (1991) describe Petri net modelling for hierarchical qualitative simulation to generate concurrent behaviour. The method of how far Petri nets support the model design during the different phases of the model development process is discussed in the paper of Kämper (1991). Furthermore, the different ways to include time in the Petri nets are evaluated.

Garnusset *et al.* (1990) discuss a simulator devoted to manufacturing systems that is based on Petri nets with objects. The simulator provides features such as: ability to define strategies of executions based on time-scheduling or priority between events; to simulate distributed systems implementations and to modify specifications dynamically.

## 7.4 Functionalities and Methodology

### 7.4.1 Functionalities

**(i) Generalities**

The simulation module will help the user in different phases of the development process of the assembly line, such as in the planning stage, the implementation stage and the operation stage. The user himself, as a whole has little knowledge of simulation science. For this reason it is very important to make an easily understandable tool available. The user should be able to carry out a simulation study without a great deal of knowledge of the methodology used to simulate the system behaviour.

The data logic embedded within the simulator will be used as a basis for the implemented production management system. This is required to extract data from the simulator tool and supply it to the on- and on/off-line modules.

The behaviour of the simulation system can be influenced, depending if one wants to validate the general system or specific management rules. Those functionalities depend on where in the development phase the user will work with the simulation module, for instance, the evaluation of the flow control rules, before deciding the layout or number of stations used is generally not very appropriate. Consequently, the necessary functions are described depending on the state of the assembly line evolution progress. Nevertheless, all the functions in all phases will always be available.

The simulator has the following two main functions:

- To supply the user with information about the dynamic behaviour of the system such as
  — the proof of the viability of the assembly system;
  — essential points for possible bottlenecks and,
  — to receive a predicted process;
- to act as a test bed for the on-line modules in order to
  — test the flow control rules and strategies and
  — to get statements for possible scheduling strategies.

The simulation module in the CAD method for industrial assembly is an essential module in the concurrent engineering scenario. It represents the link between the off and on-line modules.

# The Simulation Module

The simulation module must be able to simulate the different user scenarios. It has to be able to read the physical layout information and transform them into a suitable simulation model. The functionalities, which are mainly related to the result of on-line modules are:

- To be able to simulate with the information obtained by the other modules;
- To allow the user to interact during a simulation run.

Simulation is a useful way to concurrently test the actual control of the planned system before implementing the real process. Therefore, it accesses the local flow control rules.

Depending on the local rules, the simulated switches carry out the same decisions, when a part arrives in the same way as when a part arrives in the planned system. This allows verification of the local rules before implementing them in the programmable logical controllers (PLC).

The simulation module requires information about the planned production of each resource. It obtains this information by accessing the bills of work (BOW) created by the flow-control module.

Furthermore, the simulation module will act as a dynamic test bed for the on-line control software. The on-line modules have access to the system state as created by the simulation module. The flow control module is able to carry out global decisions depending on the system state. They will result in a global flow control rule and/or change of BOW. In summary the simulation module has to:

- access the BOW of station, feeder and exit;
- take global flow control decisions into account;
- enable asynchronous polling of the resources;
- supply the Control modules with the system state information;
- take the local routing decisions into account and validate them before they are implemented in the real system (local functionality's level);
- react on interference by the global decisions.

Figure 7.2 shows the functional scheme of the simulation module. It receives data from the resource planner and from the two on-line control modules, flow control and scheduling. The Input data and possible simulation outputs are shown here.

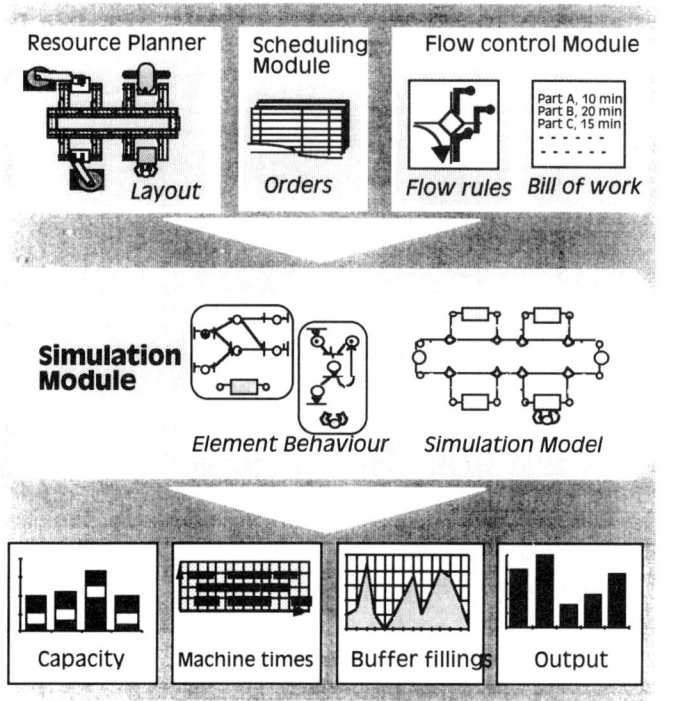

**Figure 7.2** The functional scheme of the simulation module

**Planning and conception phases** With the help of the simulation model, information about the dynamic behaviour of the production system can be achieved at the planning stage. The viability of the planned assembly system can be proved, a predicted process will be received and potential bottlenecks can be detected.

In the planning phase, the general system architecture has to be validated. In the case of job shop manufacturing, the number of workers needed and their qualification is validated using the simulation module. Furthermore different scenarios can be simulated, for instance one, two or three shifts.

In automated manufacturing, the general system architecture will be studied. Here, the opportunity to validate the layout of an assembly line is important, which means having stations in derivation or on-line. Hence, the type of material transport between the assembly cells is an important factor in the system architecture. The simulation module will give the opportunity to change the type of material transport between the different stations, for instance to use an AGV instead of the conveyor or change the length of a conveyor. Workers play an important role in automated manufacturing. Workers here are used to carry out manual assemblies, to transport material between the cells, to transport material to the cells, to carry out quality control operations or to work as systems operator. The simulator has the functionality to validate the chosen number of workers in his system to validate. Information about the qualification of the different workers and the type of work organisation to be studied.

Furthermore, scheduling algorithms, flow control rules and error recovery strategies can be tested. The logical layout of the planned assembly will be automatically received from the resource planner and physical layout editor. It can be added with default flow control rules and after launching a production order the simulation can be started (Figure 7.3).

The following functions are available:

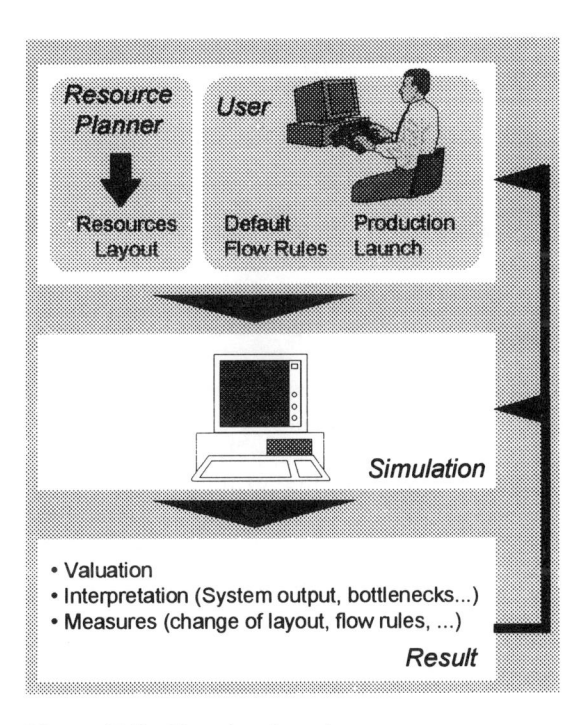

**Figure 7.3** The planning phase

*Functionalities related to the Off-line modules:*

- Evaluation of the system architecture:
  — number of stations;
  — layout (on-line, derivation, stand alone);
- change the behaviour of a resource:
  — change risk of breakdown;
  — change risk of jamming;
  — change the cycle time of the resource by accessing the resources BOW;
  — change set-up time when changing a batch;
- change of buffer size;
- simulate station breakdown;
- simulate station jams;
- Workers:
  — change number of workers;
  — give each worker different working times;
  — give each worker different qualification;
  — define and change the workplaces of each worker;
- change capacity of feeders and exits.

*Functionalities related to the On-line modules:*

- Define default flow control rules:
  — add new rule;
  — remove a rule;
- launch a manufacturing order:
  — change the order sequence;
  — change of batch size;
  — simulate urgent customer demand.

**Implementation phase** In this phase, simulation is a useful way to test the actual control of the planned system before implementing the real process. The layout of the workshop is fixed and obtained from the off-line modules (see Figure 7.4). The simulation module receives the switching rules for each switch and the bill of work for each resource. The user has the possibility to validate those decisions from the on-line modules dynamically.

Furthermore, the simulation module will act as a test bed for the on-line control software. The on-line modules will be supplied with events from the simulation module to enable the test and evaluation of the flow-control and scheduling modules,

# The Simulation Module

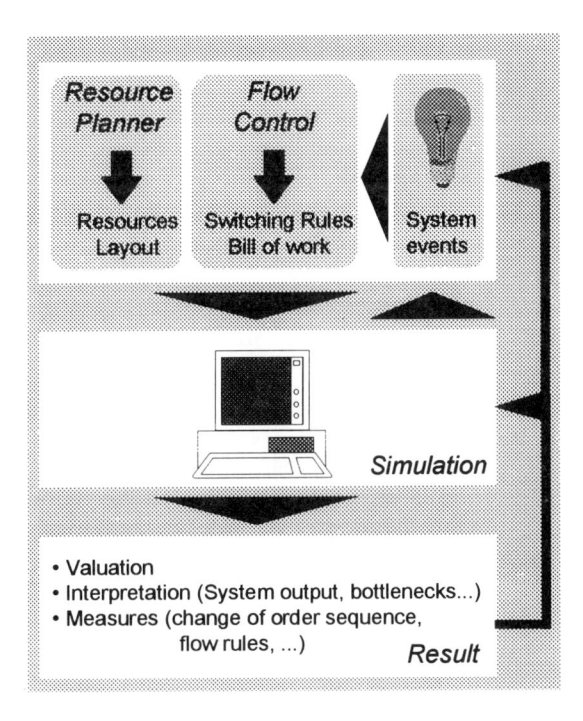

**Figure 7.4** The implementation phase

that is for those two modules the simulation module will simulate the behaviour of the real system. This means an exchange of information in real time.

During the simulation run, the simulation module and the flow control module will communicate by the exchange of events. The simulation module will simulate the real system behaviour for the flow control module in order to test the flow-control decisions. Therefore it will create events in the case of arrival/departure of parts in the case of a status change of resources. In order to have the actual production state of the simulated elements, the flow control has the possibility to poll the elements. This is be done by placing a so-called polling event in the event queue of the simulator. The simulator reacts by transferring the actual state of a specified element (or the overall system) to the flow-control model. Depending on the state, the flow-control module will react by downloading new set of bill of works (including modified targets) or a new set of flow-control rules.

In the implementation phase, the following functions of the simulator will be needed:

*Functionalities related to the off-line modules:*

- Change the behaviour of a resource (station, buffer, . . .)
  — change risk of breakdown;
  — change risk of jamming;
  — change the cycle time of the resource by accessing the resources BOW;
  — change set-up time when changing a batch.

*Functionalities related to the on-line modules:*

- Evaluate and change the flow control rules;
- simulate breakdown situations in order to test the reaction and decision of the flow control module.

**(ii) Interaction with other modules**

The simulation module represents the link between the off- and on-line modules of our CAD method for industrial assembly. The connection with other modules is described in Section 7.1. Since the integration of the simulation module in the overall system is a very important topic in the development, the interfaces to the off- and on-line modules are detailed in the following subsection.

**Interface to the off-line modules** Figure 7.5 shows the data exchange between the off-line modules and the simulation module. The simulation module uses models of the cell and station as input to obtain information about the general layout of the planned assembly system.

The data is generated by the resource planner and enhanced by the user via the physical layout editor. This data is stored in the equipment database. From this database, the simulation module will receive the resource layout, which will be represented as a Petri net. Furthermore work plans for each station, information about workers and the behaviour of each resource (break downs, jamming, . . .) will be received.

**Interface to the on-line modules** The flow-control module performs flow-control decisions in the system, when certain events occur. This decision are based on flow-control rules. The flow-control

# The Simulation Module

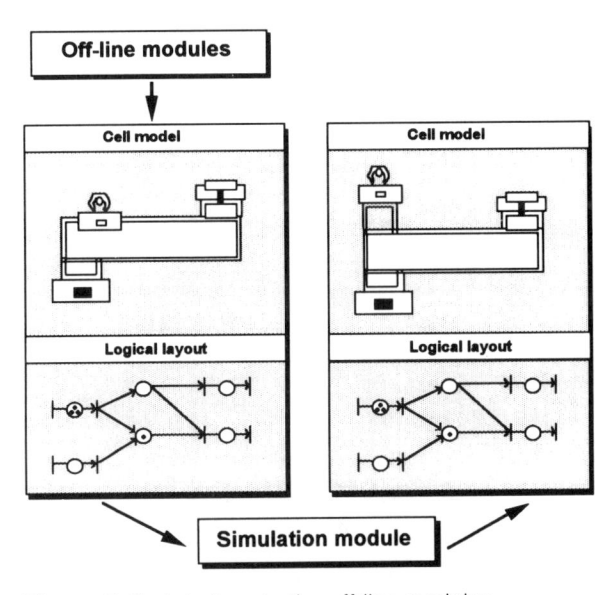

**Figure 7.5** Interface to the off-line modules

rules are distinguished as local flow-control rules and global flow-control rules. The local flow-control rules control the flow for example at one switch (only considering the input of the switch and a rule), choosing which decision to take depending on the input. A global flow control rule occurs, when global knowledge of the system state is required to choose the destination of an object that leaves a resource.

Therefore the simulation model receives for each switch in the system a flow rule. The simulation module receives flow-control rules as shown in Figure 7.6. Furthermore, the simulation module will receive the bill of work for each resource in the system.

Hence, the simulation module simulates the behaviour of the real system as a test environment for the flow-control module (see Figure 7.7). Therefore, the simulation module supplies the flow-control module with the following system events:

- arrival/departure of parts when using global rules (for example "resource A has a part to send: to which machine?");
- change of status in the resources (for example "resource A broken until date XX (if known))".

176  CAD Method for Industrial Assembly

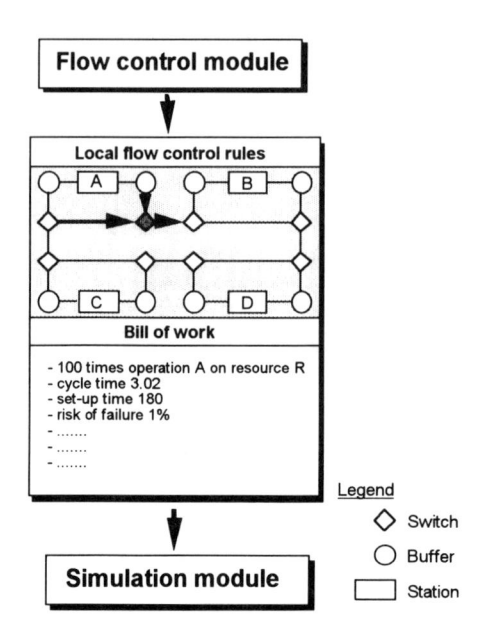

**Figure 7.6** Interface to the fow control module

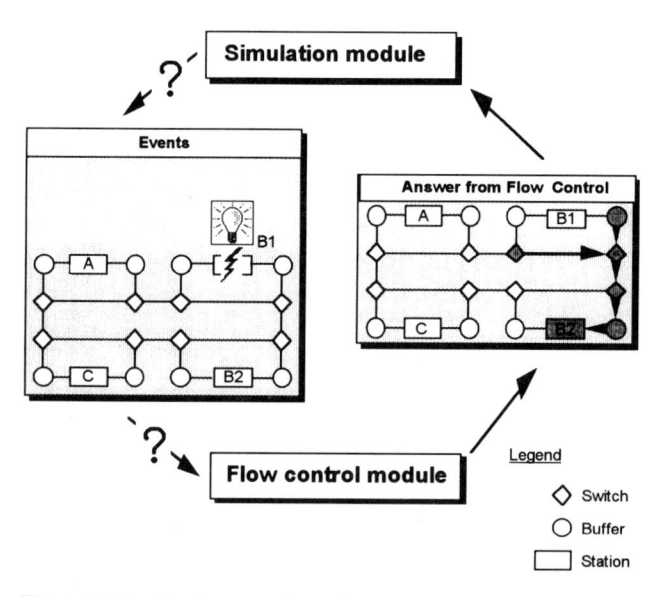

**Figure 7.7** Exchange of events

The flow-control module will then inform the simulation module with a decision, taking into account a global flow-control rule in the flow-control module. If the flow-control needs more information about the system state or the state of a specific resource, it will send a request to the simulation module to ask for this information of the global flow-control rule considering this specific event. The answer from the flow control module to the simulation module will be one of the following:

- Polling request
  "resource A: How are you, what have you done since our last meeting?";
- download new set of bill of works (including modified targets);
- download new set of flow control rules;
- answer on a global flow control event.

This will be a very sufficient way to test the flow-control software before the real system has been built.

The user has the possibility to validate the flow-control rules and to change the rules. Therefore, the simulation module supplies the flow control rules.

**(iii) Elements used in the simulator**

The simulation module models the workshop using the following elements:

- *Station*. With each station assembly, delivering, loading, repair and setting-up operations are associated. Those operations might be manual or automatic. At a manual station in a semi-automated assembly line, parts are loaded and delivered automatically, but assembled manually. Furthermore, laws about the risk of breakdowns and jams are related with a station.
- *Buffers* are specified by their characteristics such as buffer capacity, laws for jams and breakdowns if required and about the type of buffer used. Some buffer types currently specified are fifo (first-in, first-out), filo (first-in, last-out), conveyor and a random-access buffer.
- *Switches* are very interesting elements in the simulator since they are able to have various input and output links. The management of the flows is controlled by rules generated at

the flow control module. Each switch has a capacity to contain one element. A set of data related to flow control is associated with each switching node.
- *Feeders* are elements that act as a defined input into the assembly system. They allow the feeding of new parts into the system. This feeding can be done manually or automatically, depending on the required functionality. Optionally breakdown and jam laws can be associated.
- *Exits* are elements, in contrast to feeders, which acts as defined output of the assembly system and allows parts to leave the system. Like the feeders, the leaving of parts can be defined manually or automatically. Optionally breakdown and jam laws can be associated.
- *Transport-box:* A transport-box is an element that can be described as a moveable buffer or as a pallet. It can be picked-up by a transport-device and be transported to various resources via the transport-network.
- *Transport-device:* Material transported between stations not physically linked can be carried out by either an AGV or a manual trolley. To represent manual and automatic material transport system, the element transport-device is used. The difference between AGV and trolley is that the trolley requires a worker to push it.
- *Transport-nodes* represent the link of the logical transport path to the physical layout. It defines which elements will be connected via an AGV/trolley material flow.
- *Transport-network* contains information about all the transport elements and transport-paths in the model.
- *Links* describe the physical connection of elements. The links allow the flow- control module to define the paths and traffic rules.
- *Workers* are described with their performance to carry out certain operations, such as assembly or repair. This allows, for example, the modelling of a repair team or a station supervisor, which might be able to carry out all operations requires at an assembly line.

Figure 7.8 shows an example of an assembly line layout with most of the elements defined in the simulator. This layout shows an assembly line in which four stations are connected together via a conveyor system (parts or pallets being on the conveyor system are not show on the sketch). The stations are in derivation from the line. One of the stations is manual, the

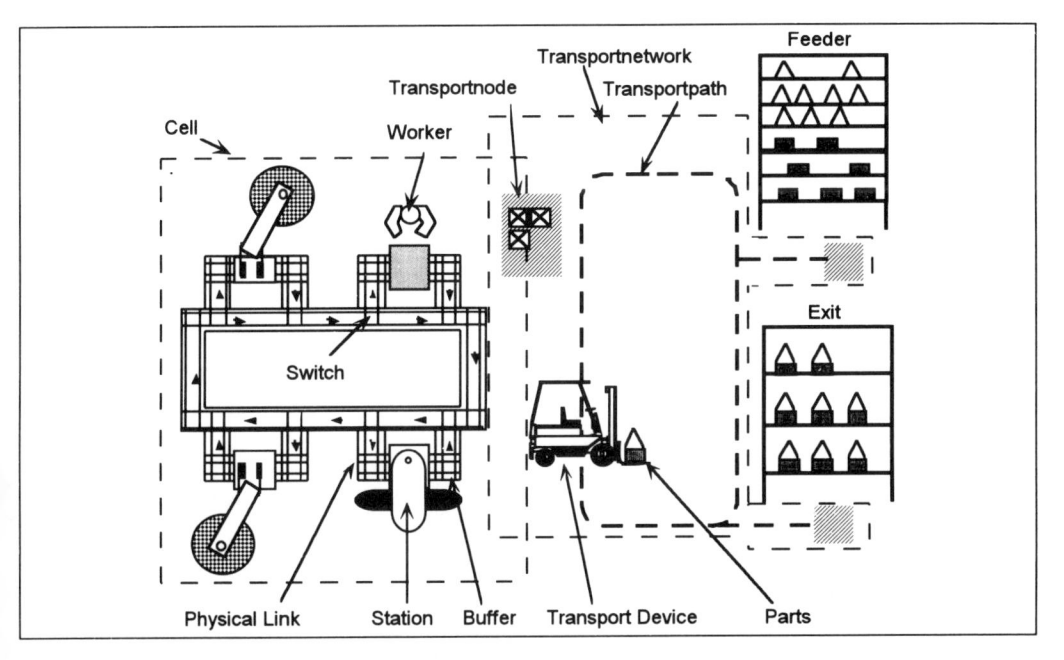

**Figure 7.8** Example of an assembly line layout, using the elements defined in the simulator

other three are automatic stations. Material is supplied with two storage systems, one contains the raw material, the other one the finished product. The manual station has to place the raw parts on the conveyor pallets and take the finished parts from the conveyor. Material transport between the line and the two warehouses (feeder, exit) will be done by a fork-lift truck.

## 7.4.2 Methodology

The development of the simulation system, focusing especially on the modelling of the elements used in the simulator, will be discussed in this section. The simulation module can be divided into two subsystems, the simulation model and the simulation engine. The simulation model manages all data and functions described in the assembly system. It contains:

- *model elements*: abstract models of the real objects;
- *time elements*: describe the "work" of the elements;
- *events*: examine the basic assembly process as an interaction of the elements (station, worker, . . .). They can be dis-

tinguished in element states, which are assigned to model elements and internal events, which are used to synchronise the different elements.

The schedule and control of the events is managed by the simulation engine.

**(i) Real system, model and simulation program**

The development of a simulation module is a procedure that consists of many single development steps. In his work a module designer is confronted with three basic elements: the real system, the model and the program. In a report published by the SCS Technical Committees (SCS, 1979) these concepts are described by

reality–conceptual model–computerised model

The conceptual model is the result of an abstraction process. The goal of this abstraction process is to select relevant system elements; their attributes and relations to one another form the variety of the given system. A conceptual model must be formally and completely described before it can be translated into a computerised model.

Hence, in order to model the real assembly process, the conceptual model is a restricted picture of the reality. This picture must be sufficient, acceptable and contain models of the elements as defined in the previous section.

Each of those elements has a certain behaviour which has to be modelled as well. This is done by defining the basic course of events for each model element. The parallel and concurrent course of these events describe the conceptual model of the real system.

To execute the simulation, the conceptual model must be implemented. A model that is implemented and executable on the computer is the computerised-model. This distinction is important because the development of the conceptual model is computer-independent. It can be implemented using different computing languages.

**(ii) Implementation of the conceptual model**

In our approach, Petri nets (Petri, 1962) are used to describe the conceptual model of the assembly scenario.

One of the ways to implement Petri nets based on a conceptual model is to use graphic Petri-net editors. The user is able to follow the dynamic change of the marks and transitions. The main application scenario is in the conception and development of control software for automated stations. However, this approach is insufficient to represent and simulate a complete assembly system, as required by our CAD method. Therefore we use Petri-nets only to describe the conceptional model.

The implementation of the model was done in a "conventional" programming language using an object-oriented approach. This means that a model for the overall system was not developed. Petri-net models were developed for each element used in the simulator, allowing the build-up of overall models out of single objects. This is a very new approach, which is based on the research result obtained by the University of Stuttgart (IAT) over the last few years (Schweizer, 1991).

**(iii) Petri-Nets**

Petri nets (PNs) are commonly used for design, simulation, performance evaluation and validation of complex distributed production systems. They are a formal declarative language simple and powerful enough to fit the requirements to model parallel and concurrent systems like flexible Manufacturing Systems and most systems in the Computer Integrated Manufacturing (CIM) environment. They are used in this case as a specification language for the simulation model because of their main features.

Here is a brief summary of the features which make PNs particularly suitable for modelling complex concurrent systems:

- Represent parallel processes and synchronisation events between the processes. This is carried out easily without the need of over-specifying the behaviour of the global system.
- PNs are non-deterministic in the sense that the sequence of firing a transition is not specified in the model.
- PN model can be easily translated into a suitable software program.

Figure 7.9 points out the correspondence between the real system and the PN model. Every event occurring in the real system can be represented as a PN transition. Whereas the different element states are represented as places, the global state of an element is marked due to transition firing.

| Real System (Production system) | Petri Net (model) |
| --- | --- |
| - Event or operation | Transition |
| - Structure | Complete net |
| - State of a unit or an object in the system | Place |
| - Global state | Marking |
| - Evolution | Transition firing |

**Figure 7.9** Correspondence between the real system and the PN model

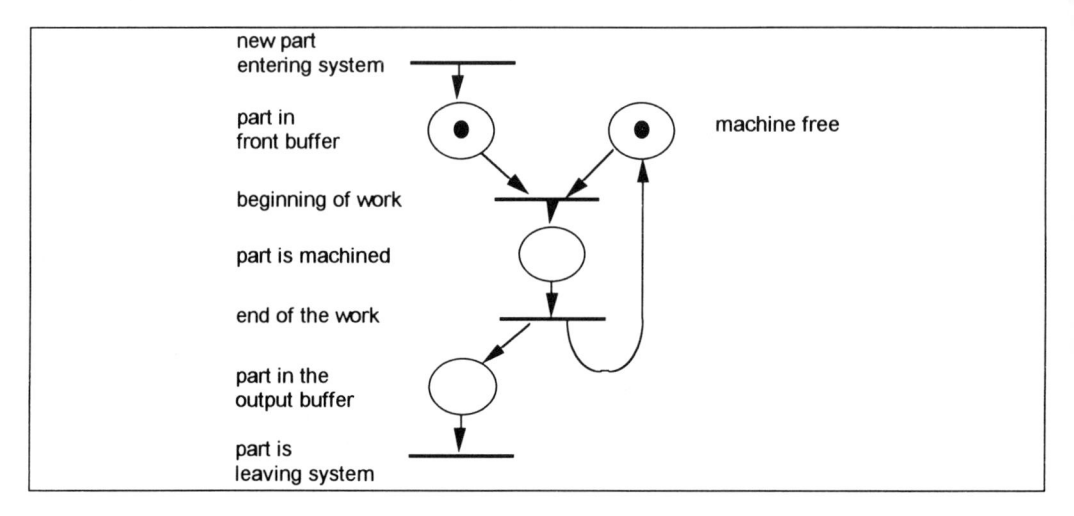

**Figure 7.10** Example for an interpreted Petri-Net

Figure 7.10 shows the sequence of a machine cycle with the corresponding buffers. The machine is able to work only if a workpiece is in its front buffer and if the machine is free. If those conditions are fulfilled, the transition "beginning of work" can be fired.

The implementation of the Petri net in program code can be carried out by implementing each transition as an independent process.

Figure 7.11 shows how a PN is able to be implemented in a program code by implementing the transition as a procedure. In this example the conditions of the places are checked before the procedure "Start Machine Procedure" is called.

The Simulation Module 183

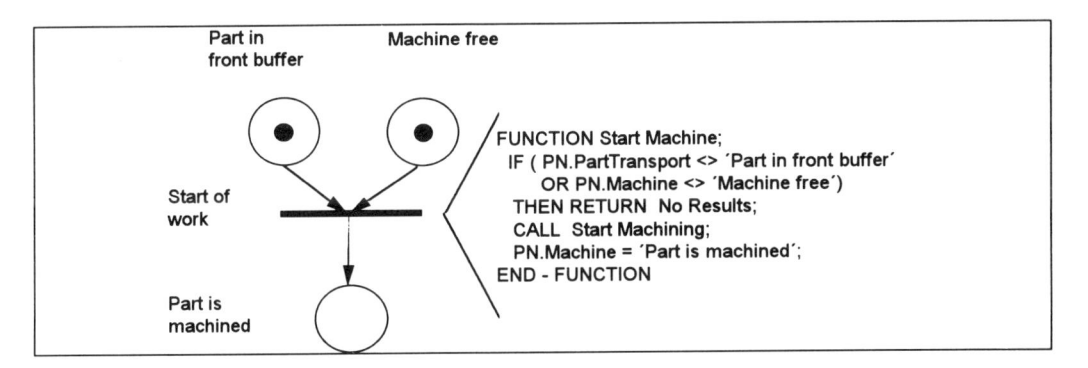

```
                Part in              Machine free
              front buffer

                   ●                      ●        FUNCTION Start Machine;
                                                     IF ( PN.PartTransport <> 'Part in front buffer'
                                                        OR PN.Machine <> 'Machine free')
       Start of                                      THEN RETURN No Results;
        work                                         CALL Start Machining;
                        ▬▬▬▬▬▬▬                      PN.Machine = 'Part is machined';
                                                   END - FUNCTION
                            │
                            ▼
        Part is
       machined            ○
```

**Figure 7.11** PN-transition implemented in program-code


**(iv) Simulation model**  In order to describe and model the behaviour of the elements in the simulator, so-called element states were defined and those events describe all the activities in the assembly system. Each model element behaviour can be described by a course of element states. Those states change dynamically during the simulation run.

Figure 7.12 shows those element states and their assignment to the model elements. Since some of these events require a certain length of time to be passed, after they occur, it is indicated in the figure which events are time dependent.

| Element states | Timed | Station | Buffer | Switch | Feeder | Exit | Transport Box | Transport Device | Link | Worker |
|---|---|---|---|---|---|---|---|---|---|---|
| receiving |  | X | X | X |  | X | X |  |  |  |
| sending | X | X | X | X | X |  | X |  |  |  |
| assembling | X | X |  |  |  |  |  |  |  |  |
| repair | X | X | X | X | X | X |  | X |  |  |
| set-up | X | X |  |  | X | X |  |  |  |  |
| arrived |  |  |  |  |  |  |  |  |  | X |
| walking |  | X |  |  |  |  |  |  |  | X |
| busy |  |  |  |  |  |  |  |  |  | X |
| not-busy |  |  |  |  |  |  |  |  |  | X |
| handover |  |  |  |  |  |  |  |  | X |  |
| get Box |  |  |  |  |  |  |  | X |  |  |
| parking |  |  |  |  |  |  | X | X |  |  |
| drive |  |  |  |  |  |  |  | X |  |  |
| on Device |  |  |  |  |  |  | X |  |  |  |
| ready to pick |  |  |  |  |  |  | X |  |  |  |

**Figure 7.12** Element states in the simulation module

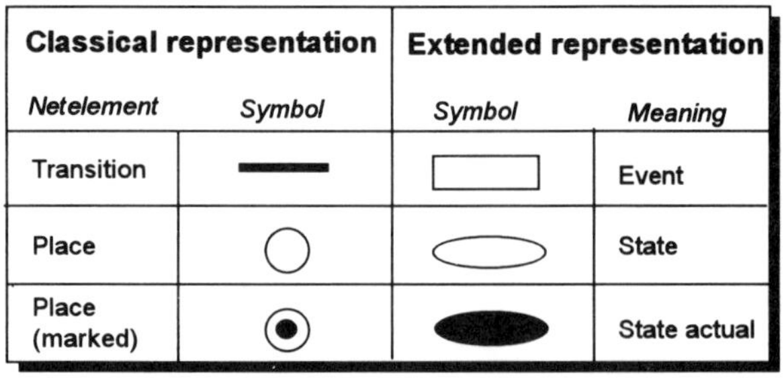

| Classical representation | | Extended representation | |
|---|---|---|---|
| Netelement | Symbol | Symbol | Meaning |
| Transition | ▬ | ▭ | Event |
| Place | ○ | ⬯ | State |
| Place (marked) | ⦿ | ⬬ | State actual |

**Figure 7.13** Changes in the traditional representation

The course of element states for each model is the basis for the Petri-net modelling of the elements. In this chapter a brief example is given as to how those Petri Nets can be obtained. In order to simplify the traditional Petri-net representation some changes in the Petri-net representation have been made (Figure 7.13). However, these changes can be easily transferred to the traditional Petri-net writing. Hence, simplifications are described in order to enable an easier way to represent certain sequences.

**Modelling of sequences** The PN of an element is built up of certain sequences. These sequences might be used by several elements, for example, the repair cycle can be found in all elements that are able to simulate faults during the simulation. They do not completely represent the PN sequence of the physical elements but are used to enable the understanding of the method as to explain how the element behaviour was modelled.

The assembly sequence is a sequence of the three element states:

- receiving (to pick up a part, automatically or manually),
- assembling (to work on the part, either automatically or manually, in simulation the corresponding assembly time has to be taken into account) and
- sending (to bring the part to the next buffer or station).

Figure 7.14 shows the Petri-net representation of this sequence.

A repair cycle consists of the Petri-net states intact, disturbed or repair. If the transition fault is fired, the state changes from intact to disturbed. The transition start-repair changes the state to repair.

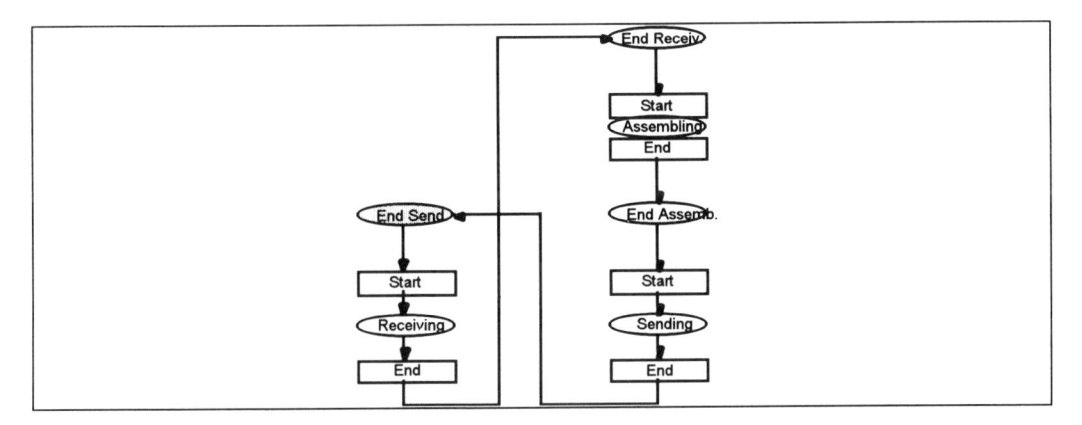

**Figure 7.14** Petri-net sequence of assembly

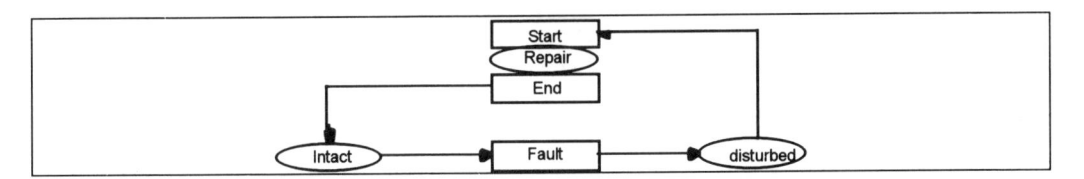

**Figure 7.15** Petri-net sequence of repair

Figure 7.16 shows a simplified Petri net for a manual and hybrid station. It consists of the repair, assembly and a set-up sequence. In order to fire a start transition, a check is made of whether a worker is present at the station (worker available). Connections to other nets, other than the worker, allow the synchronisation of elements, for example in the case of a material transport. For simplification those connections are not drawn in the figure.

**Combination of the single PN-structures to form a model system** The explained elementary Petri nets are related to one single model element. The final Petri net that is to be used for simulation will be built only when the complete assembly model is defined. This complete net has a dynamic character during simulation since the net has to be changed due to occurrences in the running system. This is especially the case for workers, who might change their places during the simulation run.

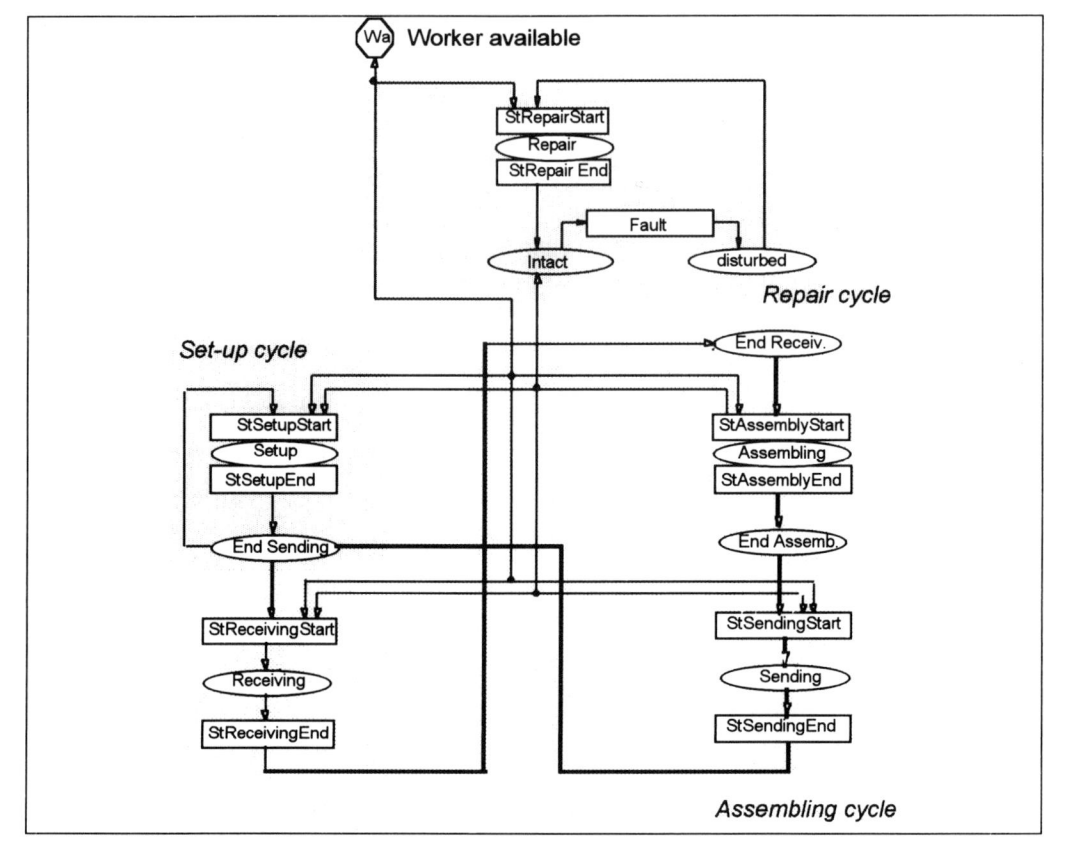

**Figure 7.16** Petri-net of a manual and hybrid station

**(v) The Simulation Engine**

The main functionalities of the simulation engine are:

- to control the chronological sequence of the state of events in the simulated system;
- to control the course of events of the simulated system.

In order to carry out those controls, the simulation engine receives information about the system state and has to take that information for its decisions into account.

The events in the simulation module are chronological, meaning that the events have to be carried out at a defined time. Therefore the event management has to "manage" the simulation time and has to synchronise the timing of the events.

That means for all elements, the same simulation clock is running. The central management of timed events is done by an event management procedure which has the following purposes:

- switching of the simulation clock;
- execution of the earliest event;
- storage of the future events (e.g. the calculated end of an assembly operation) in a chronological list;
- interruption of events in case of interruptions. The event has to be taken out of the list of future events and placed in the list of interrupted events;
- to continue interrupted events.

The modules have to react on the events that were placed in the event queue by the simulation elements or by certain simulation-engine procedures. The simulation elements place all chronological events in the queue, they are carried out when the required time is reached. Other events are events that have to be carried out immediately, such as breakdown or jam events, which are created by a special fault module.

## 7.5 Conclusions

The crucial part in any computer aided Concurrent Engineering system is to bridge the knowledge gap between the on- and off-line modules. In this chapter it is shown how a simulation module is used to integrate and validate the production planning modules as well as the control modules.

The user has the possibility to validate already at a very early planning stage the off-line module decisions, such as for example the layout of the line or the dimensioning of buffers. Concurrently he is able to supply the flow-control with a simulated system in order to test and validate the flow control rules.

The functions and the methodology of the simulator are discussed. Especially the modelling of the elements in an assembly workshop using a Petri net approach is shown. Based on the Petri-net models it is indicated how the simulator was built.

Hence, the way to test the flow control software before implementing the flow control rules in the real system is shown.

Since we have developed specifications for future computer-aided tools, the work described here might influence future developments of integrated simulators.

## 7.6 References

Antonio, E.J. and Egbelu P. (1991) Design of a synchronous manufacturing system with just-in-time production (SMS/JITP), *Proc. 13th Ann. Conf. Computers and Industrial Engineering, Ind. Eng.*, **21**(1–4), 391–394.

Boucher T.O., Jafari, M.A. and Meredith, G.A. (1990) Petri net control of an automated assembly cell, *IEEE Trans. Software Eng.*, **2**, 151–157.

Chan, F.T.S. and Smith, A.M. (1992) Simulation approach to assembly line modification: A case study, *J. Manufacturing Syst.*, **12**(3), 239–245.

Chengalvarayan, G. and Parker, S. (1991) Simulation analysis of just-in-time feasibility in a manufacturing environment, *Proc. 13th Ann. Conf. Computers and Industrial Engineering*, Computers Ind. Eng. **21**,(1–4), 303–306.

Datin, X. and Bovone, M. (1993) *Specification of user requirements*, SCOPES Project Deliverable No 3, June.

D'Souza, K. A. and Khator, S. B. (1994) A survey of Petri net applications in modeling control for automated manufacturing systems, *Computers in Industry*, **24**,(1), 5–16.

Eversheim, W., Fuhlbrügge, M. and Dobberstein, M. (1994) Simulation unterstützt die Produktionssystemplanung, *VDI-Z*, **136**, 53–56.

Garnousset, H.E., Faries, J.-F., Cantu, E., Celso, A.A. and Kaestner, A. A. (1994) A manufacturing system simulator based on the Petri net with object model, modeling and simulation, *Proc. 1990 European Simulation Multiconference*, SCS Europe, Ghent, Belgium.

Hatono, I., Yamagata, K. and Tamura, H. (1991) Modeling and on-line scheduling of flexible manufacturing systems using stochastic Petri net, *IEEE Trans. Software Eng.*, **17**(2), 126–132.

Heger, R. (1994) Use of simulation in a concurrent engineering scenario, *European Simulation Symposium 94*, Istanbul, 9–12 October.

Heger, R. and Richter, M. (1994) Modelling and Simulation in Concurrent Engineering, *ESPRIT-CIME 10th Annual Conference*, Copenhagen, 5–7 October.

Kämper, S. (1991) On the appropriateness of Petri nets in model building and simulation, *Syst. Anal. Model. Simul*, **8**, 689–714.

Okada, K. and Ushio, T. (1991) Petri net based hierarchical qualitative simulation and its application to a co-generated plant, *Application and Theory of Petri Nets, 12th Int. Conf.*, Univ. Aarhus, Denmark.

Petri, C.A. (1962) *Kommunikation mit Automaten*, Dissertation, Bonn University.

Schweizer, W. (1991) *Entwicklung eines interaktiven Simulators auf der Basis von Petri-Netzen zur Modellierung und Bewertung hybrider Montagestrukturen*, Berlin, Springer, 1992, zugl. Dissertation, Stuttgart University.

SCS Technical Committees (1979) Terminology for model credibility, *Simulation*, March.

Spur, G., Krause, F.-L. and Mertins, K. (1993) Simulationstechnik für Produktentwicklung und Planung, *ZwF*, **88**, 295–301.

Teng, S.-H. and Black, J.T. (1990) Cellular manufacturing systems modeling: The Petri net approach, *J. Manuf. Syst.*, **9**(1), 45–54.

Toncich, J. (1992) Multilevel simulation for advanced manufacturing systems, *Int. J. Adv. Manuf. Technol.* **7**, 178–185.

Wiendahl, H.-P. and Scholtissek, P. (1993) Produktionssimulation als Versuchsstand für Fertigungsstrukturen und PPS-Verfahren, *VDI-Z*, **135**(3).

# 8 *The Scheduling Module*

## *8.1 Introduction*

The scheduling module of the CAD method for industrial assembly is an automated scheduling aid that proposes schedules to the user, and was designed to answer to the specific needs of a concurrent engineering approach.

The module schedules customer demands or production orders that were planned by the production planning system, instantiates them (creates bills of work for the resources that will process the orders) and provides them for the flow-control module for order-release and down-loading.

It is situated at the factory floor or workshop level, and is designed to interact with existing factory planning systems that take care of the planning and possibly predictive scheduling phases. If no global planning system is required, the module can equally schedule customer demands directly (see Figure 8.1).

The module has been designed to react to events that arrive on the factory floor in real time (reactive scheduling), by providing a constantly updated schedule (at will) to the user, and to allow an easy interaction with him or her, through the provision of user-friendly functionalities.

The module replies to real-time scheduling needs expressed by the industrial partners. The main goal here was to provide a tool that, while being closely linked to current and future needs in industry, takes into account the technological constraints and the culture of the modern production environment.

The integration of the module within the architecture (see Figure 8.2), described in Chapter 3, promotes efficient real-time capabilities through the interaction with other modules (Anon, 1994).

The module is present in the simulation and control levels of

**Predictive Scheduling**

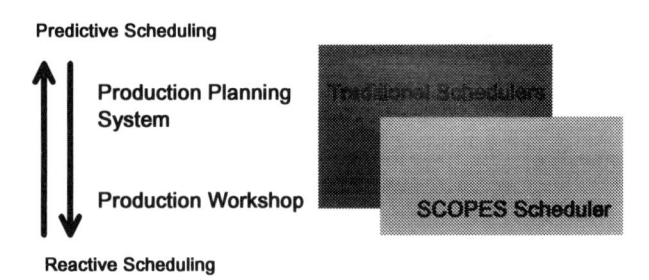

**Figure 8.1** Situation of the scheduler with regards to traditional schedulers

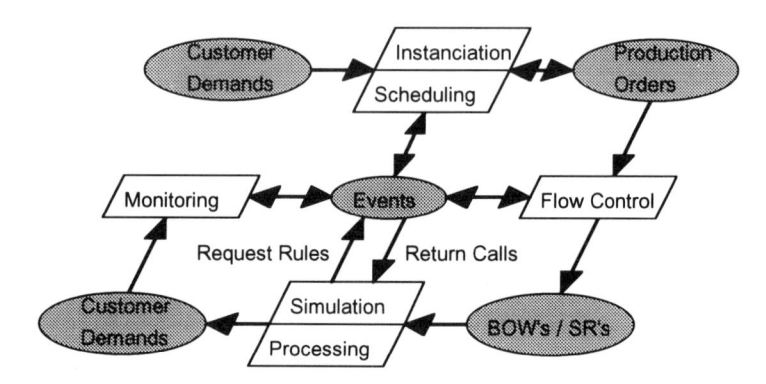

**Figure 8.2** Situation of the module within the architecture of the CAD method for industrial assembly (BOW: Bill of Work; SR: switching rule)

the architecture presented in Chapter 3. Figure 8.2 summarises the two superposed levels. Each level is inspired from the control loop based on the cubic circuit in Chapter 3. As can be seen from Figure 8.2, the two levels have essentially the same structure in both the on-line and off-line parts, which are presented further on.

## 8.2 State of the Art

In today's production environments, scheduling is a tool that is typically used during the final stages of production planning. The resulting schedule is then sent down to the factory floor, where it is realised. Any unforeseen events that arise either from the later stages of production planning or from the factory floor cannot be taken into account.

In certain environments, scheduling is performed on the

factory floor, but this is often done manually, by the factory foreman, without any aid from modern scheduling tools.

Advanced scheduling systems are only present in a few very specific environments. These systems are application specific, often being only applicable to the system they were designed for. It is therefore very difficult to provide a scheduling system adaptable to all users' needs.

Recent market forces and their resulting management and production strategies are decreasing production batch size, and increasing the number of variants of products. These tendencies are being felt directly at the workshop level, where the workshop foreman has an ever increasing amount of different batches of smaller sizes to schedule during the day, with the complications that this creates (Van Dam et al., 1993). Scheduling is traditionally done manually, as stated before, but this function is becoming more and more difficult. Unforeseen events also complicate the picture, leading to further difficulties for the foreman. The provision of a real-time workshop scheduler is thus quickly becoming an important requirement in modern workshops (Taylor, 1990).

Recent developments in supervisor and system hardware have much improved higher-level supervision and control. Quite simply, more information is available, allowing improved supervision of product batches in real time, and improved compensation strategies for the production line (Verdebout, 1992). These developments make the workshop more flexible and adapted to the trend in decreasing product batch sizes, but increase the requirements on scheduling tools.

New approaches to system supervision are also being currently applied, especially in the direction of distributed supervision architecture (Lauvigne, 1994). They result from a better understanding of the needs of modern production environments, and of the advantages of these new supervision methods. Scheduling is not yet an integral part of concurrent engineering approaches or traditional product development methods. This is particularly felt at the conception level, where the influence, and the choice of the right scheduling strategy can be an important part of the definition and the optimisation of the production layout.

## 8.3 Users' Needs

The main requirement of this module in a concurrent engineering environment is to allow the integration of scheduling, equally in on-line applications as well as simulation, to increase the global efficiency of production and simulation, thus making our concurrent engineering approach more efficient.

The elaboration of the module was supported by a precise questioning of interested users, to increase its utility. As a result, the module replies to current market tendencies.

One of the significant results of questionnaires aimed at industrial users was the need for an aid for scheduling at the workshop level, as opposed to most scheduling systems that are part of the production planning level.

The majority of the users equally insisted that the tool cannot replace the operator through the provision of fully automated scheduling. In other words, the operators require a computed schedule that they can then validate or first modify, to add additional constraints that cannot be taken into account by the system. This approach allows a simplification of the model by decreasing the need for accuracy, and greatly increases user-friendliness.

These requirements and others defined by the industrial users were taken into account; the most important ones are listed here below:

- *Users' requirements for global scheduling:*
  — Most of the users require global scheduling.
  — For most of the users, the limiting factors are the availability of operators, resources and components.
- *Users' requirements for local scheduling:*
  — Most of the users do not perform local scheduling, but would benefit from it.
  — For most of the users, the limiting factors are the availability of operators, resources and components.

Error recovery through rescheduling is deemed necessary by the users.

The scheduling module, as defined in the CAD method for industrial assembly applies to the following scenarios:

- mid-size to mass production volumes, fully automated or hybrid systems;
- fast just-in-time assembly systems, even with a lot size of one product per subassembly;
- manual assembly of small series is not foreseen.

In the mid-size to mass production scenarios, MRP, MRP II management systems are used to good effect. Global scheduling is to be foreseen as a rescheduling tool. Local scheduling is more applicable to the job shop scenario: as the permutation of batches is not possible within an automated permutation flow shop transfer system (permutation flow shops make up the vast majority of automated production flow shop systems, and in particular, assembly systems, thus implying the more frequent use of input sequencing in a global manner, or rather global scheduling). The cycle time here is of the order of a second. Batch control may or may not be implemented, according to the users' needs.

In just-in-time assembly scenarios, no scheduling by the module is required, as the product flow is, by definition, pulled by the demands. The cycle time here is of the order of a minute. The batch size being naturally equal to one, there is no need for its control.

This module, in close interaction with the other modules, not only takes advantage of new developments in supervision system hardware and control methods, but more importantly, it takes into account current industrial culture, and in particular, the working habits of workshop operators.

## 8.4 Functionalities and Methodology

### 8.4.1 Characteristics

**(i) Interaction with other modules**

The scheduling module is present in both the off-line and the on-line parts of our concurrent engineering approach.

Off-line, the module interacts with the other off-line modules, aiding in the concurrent development of the product and production means. It is implemented together with the other off-line modules. The role of the module is to aid in the choice of the topology of the installation and in the choice of the right scheduling method. This evaluation is carried out through simulation (see Figure 8.3).

The choice of the topology is carried out through the interaction of the scheduling module with the flow-control (Saunders et al., 1990) and simulation modules, via the definition of the model of the installation in which the product is to be

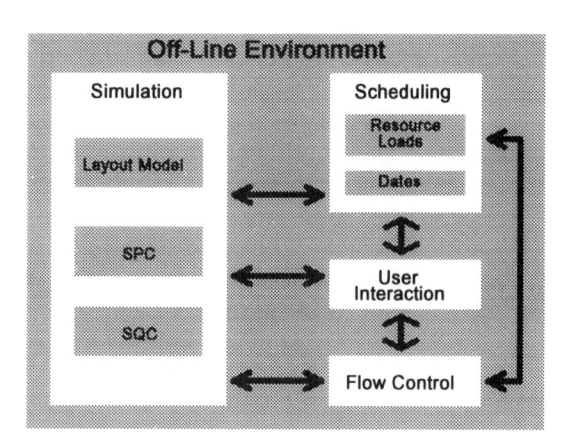

**Figure 8.3** Major off-line scheduling module interactions

produced, and the flow-control and scheduling strategies. The sensibility of the system topology with regards to the choice of scheduling method can thus be established.

The choice of the scheduling method is equally done at this stage. Many important indicators are brought to light here, such as the ability of the method to react to different surprises, its average performance, and other phenomena that may not be visible at first analysis.

On-line, the module interacts with the other on-line modules, providing a schedule in real time and providing the user with reactive scheduling capabilities. The module is implemented in the supervisor, with the other on-line modules (see Figure 8.4).

**(ii) General functionalities**

The essential characteristics and functionalities of the module are the same both off-line and on-line, including the user interface, which has the same physical aspect on the CAD station and the supervisor (see Section 10.5). This means that both the product development team and the production line operators are clear as to the way the module is to be used.

The same scheduling strategies are available both off-line and on-line, allowing a direct transferring of off-line conception results to the shop floor supervisor. This is possible through the definition of a standardised format, which also allows the implementation of user-defined heuristics, if an application-specific strategy is required. These application-specific heuristics

# The Scheduling Module

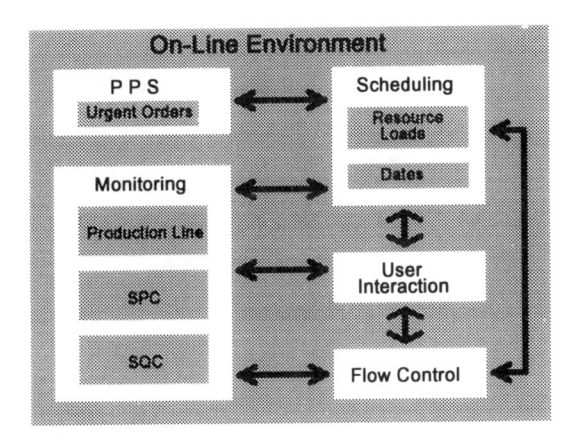

**Figure 8.4** Major on-line scheduling module interactions

can be implemented in an independent adapted computer that is connected to the supervisor, to avoid the overloading of the supervisor and decreasing production performance. This approach also replies to the Achilles' heel of scheduling methods in general, namely the general inadaptability to problems that are different from the one they were designed for.

This module is thus an integral part of the concurrent engineering approach, and is equally one of its primary distinguishing features. Global system evaluation and configuration is not only possible, but systematically applied, to take into account all the important constraints of the production environment.

On-line, the module provides the user with a list of the same simple heuristics for real-time scheduling of the line. This choice has been voluntarily limited, as the module is implemented in a supervisor with limited computational capabilities, and is required to reschedule in a short time span.

Furthermore, the module has two distinct scheduling modes, that have been called global and local scheduling, respectively (see Figure 8.5).

Global scheduling allows the scheduling of the workshop in a simplified manner, to give a general idea of the performance of the general schedule of the workshop. If used in a real-time manner, it allows a fast estimation of the quality of the schedules of the production orders in the workshop, of the load of the cells, and a rough estimation of the finishing dates of the orders. If no higher-level planning system is used, it is this

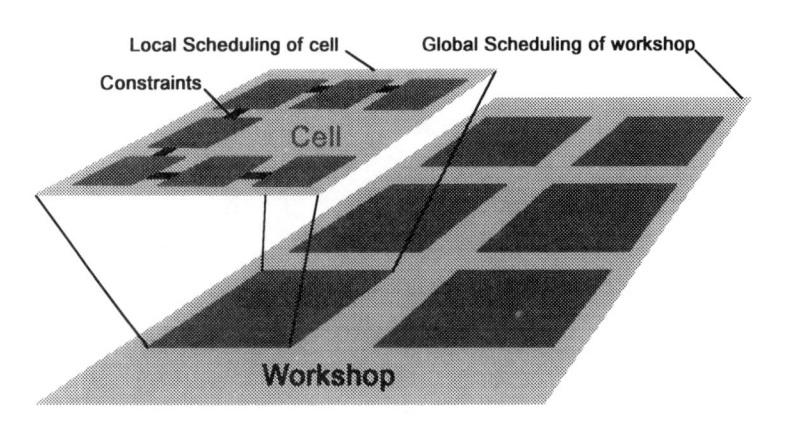

**Figure 8.5** Global and local scheduling levels

functionality that provides general system scheduling (note that we differentiate between planning and scheduling, the former being, by definition, an estimation of future requirements, and the latter being the actual production to reply to real demands).

Local scheduling allows a more precise scheduling of the workshop, taking into account all the constraints that are present in the representative model. This level allows the precise computation of the schedule, taking into all the constraints and details of the production-line model data (the model is either that of the simulation, off-line, or a billboard model with production line information, on-line).

These two modes give the module increased scheduling functionalities, flexibility and autonomy, as no global predictive scheduling system is required for its functioning. Global scheduling allows infinite loading, as the global loads of the cells can be computed directly, without taking into account the assignments of the resources, as with a traditional infinite loading scheduling system. Local scheduling allows finite loading, with all the modelled constraints being taken into account. The user is simply free to use the module at the level of detail and performance he requires.

If more precision is required (this is especially the case during on-line production), due to constraints that are impossible to represent in the model, the user can add in his own in real time and evaluate their influence on the system.

With global scheduling, the following functionalities have been provided for:

- capacity requirements planning;
- launching.

Capacity requirements planning allows the computation of the general loads on the resources, directly from the customer demands, if required. It is divided into the following functionalities:

- stacked lead time;
- load;
- capacity smoothing.

The user can deduce the general loads for the concerned scheduling period by first introducing the production orders in the order that they were received, and then evaluate their effect on the system. He can then proceed by scheduling the orders to balance the system.

Launching works together with the flow-control module. The orders and jobs are made available for launching by the flow-control module, before the minimal launching period has expired.

Local scheduling being situated at the resource level, it receives the bills of work destined for the resources, and reschedules them according to the chosen scheduling heuristic. It then proposes the modified schedule to the user, who then either confirms it, modifies it, or keeps the old schedule.

Local scheduling will typically be done to optimise one of the following situations:

- When time is of the utmost importance, scheduling will consist of job sequencing in order to reduce the total travel time.
- For bottleneck resources, the scheduler will aim at the reduction of set-up times, and all the other unproductive times.
- When resource running costs are high, the orders will be sequenced with this in mind.

The following functionalities have been met by the module:

- The two categories of users (on and off-line) can directly use the results of simulation to aid in the scheduling of a real system. This point denotes our concurrent engineering approach.

- The customer demands or production orders are sorted to optimise cost constraints, while taking into account mandatory system constraints.
- The module presents a list of bills of work to the flow-control module.
- A set of standard scheduling strategies has been provided within the module.
- The module can apply user-defined scheduling methods, provided that these methods respect the calling-and-return format specified for the module.
- The module allows error recovery through rescheduling.
- The production orders are in communication with the customer demands through a many-to-many link, which allows their grouping and splitting.
- The production orders are equally linked to the bills of work, and it should be noted that these bills of work are themselves directly related to the bills of materials of the process. This implies links between different modules, and a corresponding support system. In other words, as a functionality, this module provides a support for this support system.

As this module is related to both on-line applications in a supervisor and off-line applications within the simulator, different goals were aimed at.

The following constraints have been taken into account:

- the limitations and constraints of current scheduling systems and their technological and practical constraints;
- the technological limitations of current controllers and in particular, their programming techniques;
- the technological limitations of current supervisors and their particular programming techniques.

The man–machine interface (MMI) provided with this module allows the user to easily access the production orders to be scheduled, as well as their corresponding customer demands. The main interactions of the user are the validation of a proposed schedule, the modification of an existing one, or the modification of production orders, such as grouping, inserting, removing, etc.

The user is able to perform a fast what-if analysis of the effects of modifications to the schedule, either by the user, or the module. At the workshop level, this is performed by global scheduling, and at the resource level, by local scheduling. The what-if analysis consists essentially of modifying the due dates, simulating breakdowns, and taking into account the effects of the modifications of other system parameters, such as the assignment of resources, etc.

Parts of the module being closely linked to the shop floor via the other modules, it is able to do a finer scheduling and error recovery through rescheduling, both in its precision and time period. It is through these communication links that it is able to react quickly to any unforeseen events, and propose a modified schedule that takes them into account.

Error recovery is performed by the module, in collaboration with the flow-control module, which may create new batches to compensate for losses during production, or possibly by modifying routing strategies, etc. Error recovery works roughly as follows.

If a resource or part of the production line is stopped (an extended period of time, defined by the flow-control module) for any particular reason, the scheduling module may receive this information (called an event) from the flow-control module, which is in charge of error recovery and decides upon the gravity of the event. The scheduling module may then be prompted to change its schedule to take the implications of the event into account, after the notification of the operator. The operator then decides upon whether a new schedule is needed, or whether to simply keep or modify the existing one. He or she then validates the new schedule, even before the work remaining on the resources is finished. The module is thus able to react to events at the resource, cell and workshop levels.

The processing of urgent orders is possible, and is facilitated by the provision of numerous user-friendly functionalities, such as the splitting of the currently processed order.

The scheduling module is, in other words, a real-time scheduling module that reacts to any events from the user or from the system, especially those associated with error recovery, thus complementing the traditional MRP style predictive scheduling modules.

### 8.4.2 Methodology

The scheduling activity usually starts off during the conception of the assembly line. During this phase, all the off-line modules are in the workstation used to elaborate the product and its production means. Once product design has sufficiently advanced to be able to define the resources and the layout, the scheduling module comes into use. The usual work method is to define the simulation model of the layout, followed by its chosen flow-control strategies (also to be evaluated). The scheduling methods to be evaluated are then implemented within this model. In this manner, a global simulation model, which takes into account the main technological constraints and limitations of a real production environment, is defined.

This simulation model permits the simulation of real and proposed installations over a prolonged period of time, to take into account the effects of global and local scheduling in a more precise workshop simulation. Furthermore, the interactions of a new product with already existing ones on the production line can be assessed.

When moving from simulation to the real world, the chosen methods and parameters are downloaded to the real system, the results being then directly exploited within the real system. This is possible as the concerned modules have the same structure both off-line and on-line.

Being an integral part of the architecture defined in Chapter 3, this module has several notable differences to traditional schedulers. A scheduler usually starts off with customer demands and produces sorted production orders; however, this is not always the case. When this module is used in conjunction with an MRP style production planning system, the customer demands have already been instantiated into production orders, which are the starting point of the module. In this case, either global scheduling is not performed by this module, only local scheduling, or global scheduling is still performed to further optimise the schedule if the need arises.

Given a set of customer demands and a system containing a set of physical resources that are able to satisfy these demands, the scheduling activity consists of two approaches:

- defining in which sequence these demands will be launched in the workshop and its constitutive cells—this is called global scheduling, and corresponds to input sequencing;
- defining in which sequence the jobs of an assembly sequence will be launched in the machines (or manual stations)—this is called local scheduling.

The following basic tools were specified, to allow real-time scheduling of the workshop:

- main functions of the module;
- infinite loading;
- finite loading;
- basic scheduling strategies.

With these basic tools, capacity requirements planning and launching (with the interaction of the flow-control module) are provided for.

The chosen on-line architecture is of the three-tier type, namely station–cell–workshop. On-line, this module is placed at the workshop supervision level. The database models being recursive, the module uses this property to its advantage. The characteristics of global and local scheduling are essentially the same, as the same functionalities are available for both global and local scheduling.

Because the database structure is recursive, each of the subcells has the same structure as the main cell it is part of. This will essentially impose a recursive approach to the scheduling of the cell or resources. Thus the global scheduling of a cell of the level below will be the local scheduling of the cell defining the level just above (see Figure 8.6).

It should be noted that the scheduling activity is vastly enhanced by being based on our concurrent engineering method's databases, which are closely linked to the representation of products and their production means by means of processes. With these databases, it is far easier to construct a

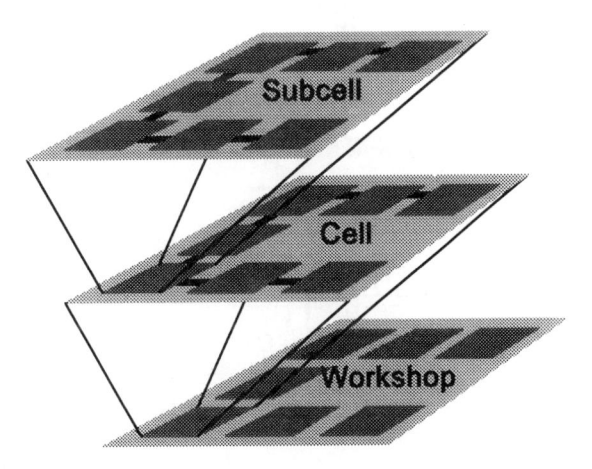

**Figure 8.6** Recursive structure of the database model

precise model for schedule computations, including all the important production line constraints.

Another advantage of our database structure is its object-oriented approach, which in the case of scheduling means that a different scheduling strategy can be used for each specific resource, thus vastly extending the applicability and adaptability of the module.

This approach also simplifies the specification and implementation of the tools, by allowing the specification of common functions. Furthermore, the infinite loading and finite loading methodology are very similar, as a result of the database.

In essence, the main difference between the two types of scheduling will be the accessing of the local as opposed to the global bills of work (see Figure 8.7).

In the same manner, the date and load computation procedures use the process model structure and database advantageously. The infinite and finite loading procedures use the process structure associated with the production orders to

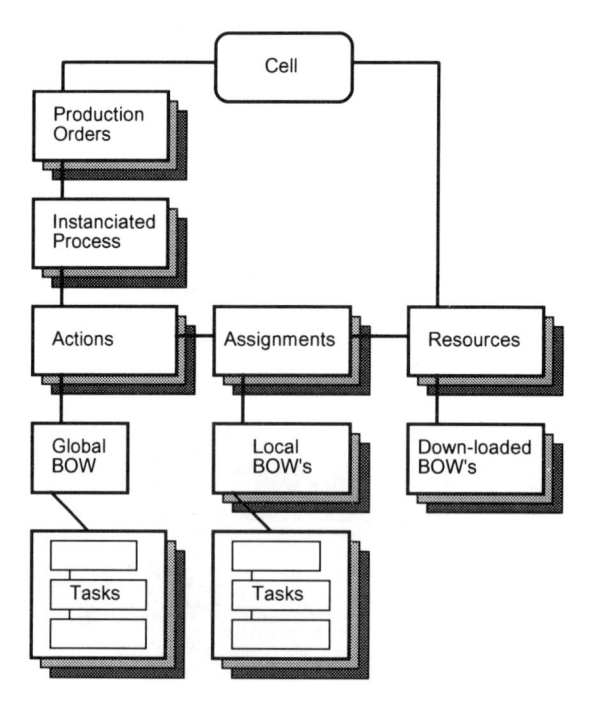

**Figure 8.7** Local and global bills of work

create a representation of the bills of work (called tasks) of the associated actions (for infinite loading) or resources (for finite loading). These tasks are different, depending upon whether the transfer mode of the buffers between the resources is of the batch or overlap transfer kind. Furthermore, the cycle time and other performance parameters of the different resources are integrated within the parameters of the tasks, to obtain more reliable estimations of the results. The different tasks are linked together with links that represent the system constraints, be they hard constraints (such as, for example, automatic transfer systems between two resources), or soft constraints (such as an ordered constraint between two permutable bills of work). A graph of tasks and links is thus constituted.

Being in direct communication with the other modules of the CAD method for industrial assembly, the scheduling module is able to access the best available information concerning the estimations of the durations of the different resources, be they transmitted by the resource planning module, or the user off-line or supplied by the monitoring module during production. This means that both off-line, and on-line, the estimations of the schedule become more and more precise with time.

The scheduling of the workshop is done as follows:

- If global scheduling is required, the production orders of the workshop or cell (represented by what we call global bills of work) are scheduled, according to the global strategy.
- If local scheduling is required, then the local bills of work that are present at the resource level are scheduled. These constraints are introduced into the graph, completing its constraints.

The dates, loads and other important parameters are then computed, based on this graph. The recursivity of our model structure means that the global bills of work of a cell in a workshop become the local bills of work at the level just above, and so on, thus allowing an unrestricted amount of levels of representation, and detail if required (see Figure 8.6).

This approach has numerous advantages, especially the easy creation of a schedule date computing method, an easy validation of the model and results, an easy representation of the different assignments, loads and other system constraints, and above all, a model that is easy to comprehend by the user.

With regard to the algorithms that were developed for the

module, except for the scheduling heuristics themselves, all the foreseen computations are non-exponential, meaning that their complexity can be compared to that of a sorting algorithm. This allows their application without the imposition of important constraints on the supervisor hardware.

As a result, these computations can be executed as many times as required, and more importantly, when they are most needed.

As far as the scheduling heuristics are concerned, this need not necessarily be the case, as is generally known from previous experience of scheduling problems. It was decided to leave this problem to the responsibility of the user, who will decide upon the best heuristic to apply, by taking into account this practical consideration. He is nevertheless aided by the other tools of the integrated CAD method in his choice.

During scheduling computations, an intermediate schedule can also be used. Generally speaking, one can divide scheduling methods into constructive and enumerative ones. The former give the best schedule through the application of an algorithm. In this case, as long as the scheduler is computing the schedule, the rest of the system has to wait. The latter methods list or enumerate all or some of the possible schedules, and eliminate those that are non-optimal. In this case, the scheduler has the best available schedule during the computation, which may or may not be the optimal one, if indeed that is possible for the problem! Therefore, if the need arises, the scheduler can give this intermediate schedule for down-loading, or at least the next order.

This is particularly advantageous in the application of artificial intelligence scheduling techniques, where the scheduling list becomes more and more optimal with time. The construction of robust schedules is equally possible, through the implementation of the right heuristic, even though this point was not directly addressed during the project.

Every occurrence of an error may result in a new computation of the dates, thus providing a powerful error-recovery tool. One of the possible methods is to compensate the loss of parts with a new production order, to be inserted later in the schedule. This point is dealt with in more detail in the flow-control chapter.

### 8.4.3 Functionalities

The following functionalities will be provided for through the interaction of this module with the other modules:

- Provide visible customer demands and production orders to the user, through the man-machine interface that maintains such lists and allows their control.
- Propose scheduling heuristics to the user.
- Provide for the definition of user-defined heuristics, for example, space constraints.
- Take into account major events from the flow-control module, and propose eventual rescheduled sequences.
- Manage error recovery through rescheduling.
- Allow for the processing of urgent orders, that may arrive in the middle of the processing of another production order. Here as well, the splitting and modification of orders will be user activated.
- Provide an algorithm for the calculation of the main dates, given a sequence of orders, and an initial situation.

The following hypotheses were made during the elaboration of the module:

- The workshop is generally of the flowshop type, even though both types can be treated by the scheduler.
- Both batch (batch transfer or batch processing) is a manufacturing operation in which a designated quantity of material is treated in a series of steps. Also a method of processing jobs so that each is completed before the next job is initialised (started) and overlap transfer (Overlap transfer or overlapped schedule is a manufacturing schedule that "overlaps" successive operations. Overlapping occurs when the completed portion of an order at one work centre is processed at one of more succeeding work centres before the pieces left behind are finished at the preceding work centre(s)) are treated.
- The resources are generally considered to be non-pre-emptive (a launched production order on a resource must be finished, before commencing with another one), except for the transfer resources.
- Only the topology of physically constructable installations will be considered, in as far as the elaboration of the task and constraint graph is considered.
- The times used in these computations are the average times supplied by the user.
- The general process structure of the CAD method for industrial assembly is applied in this module.

During the elaboration of the module, it was decided to compute the dates and loads through the application of a task and constraint graph, to allow the utilisation of common procedures between both infinite and finite loading.

### (i) Main functions of the module

The following functions have been provided to facilitate this module's utilisation:

- Add a new production order.
- Remove an existing production order.
- Shift a production order one step ahead in the sequence.
- Shift a production order one step back in the sequence.
- Split a production order into two.
- Group two production orders.
- Change the target value of production order.
- Add an assigned resource.
- Remove an assigned resource.
- Infinite loading.
- Finite loading.

This set of functions was chosen to make the module sufficiently user friendly in an industrial workshop environment, and at the same time, sufficiently small for easy development and implementation within a workshop supervisor.

### (ii) Infinite Loading

Infinite loading was provided as an aid for the manual smoothing of the load by the user. Both batch transfer and overlap transfer infinite loading were provided for.

Infinite loading was designed to be computed for a single cell level at a time, therefore, if the user wants to compute the loads of one of the subcells of the workshop, he must first place himself at the level of the subcell. This methodology greatly simplifies the procedures, whilst still allowing precise computation of all the cell levels.

As explained earlier on, the methodology for infinite loading is essentially the construction of a graph representing the jobs to be done (tasks), and the constraints between them. At this level, only the global bills of work are scheduled from the production orders and their associated processes.

Depending upon the level of detail required, the scheduler may find itself at one of numerous levels, defined by the three-tier structure. If the scheduler is at the workshop level, it schedules the orders for the workshop, without taking into account the assignment of the cells or resources. If it is at one of the cell levels, it schedules the orders for the cells at the chosen level, without taking into account the resources of the cells.

The computations are done as follows. Firstly, the production orders assigned to the workshop or cell are scanned. Then, their corresponding process graphs are scanned. Note that these process graphs have been instantiated and are no longer generic process graphs of the product type, therefore they not only contain the pertinent information as to which actions are appropriate, but also the resources to which the actions can be assigned.

The instantiation method of the process graphs is detailed in the chapter on flow control.

Whilst scanning the instantiated process graphs of the production orders, the bills of work assigned to these actions (global bills of work) are scanned and scheduled. The bills of work are represented in the task and constraint graph as tasks, and the results of scheduling, as well as of the order of the actions of the different production order's processes, as constraints.

If the transfer method between the actions is of the batch transfer kind, the bills of work are represented by a single task, which takes into account the batch size, and the estimated cycle time and estimated performance parameters of the assigned resource(s) (see Figure 8.8).

If the transfer method between the actions is of the overlap transfer kind, the bills of work are represented by a variable number of tasks, varying from one to three, that take into

Actions and their global bills of work    Associated Tasks

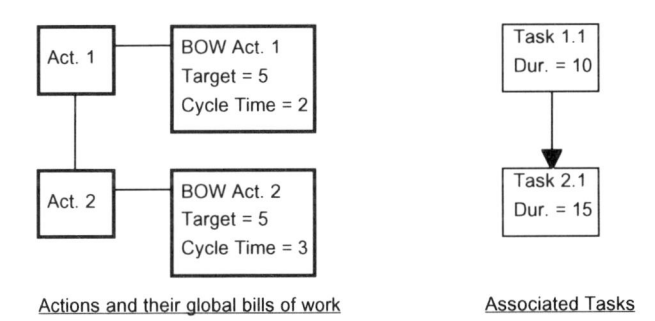

**Figure 8.8** Infinite loading batch transfer task and constraint graph

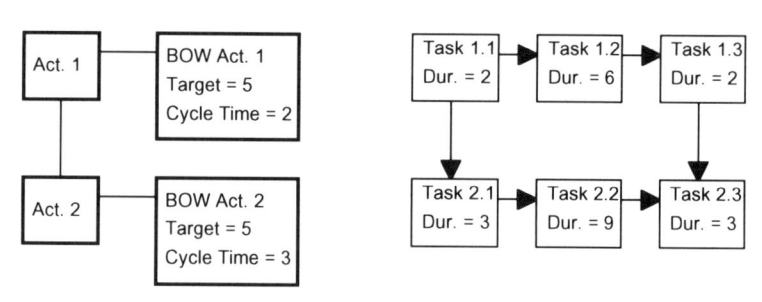

```
┌────────┐    ┌──────────────┐         ┌────────┐  ┌────────┐  ┌────────┐
│ Act. 1 │────│ BOW Act. 1   │         │Task 1.1│→ │Task 1.2│→ │Task 1.3│
│        │    │ Target = 5   │         │Dur. = 2│  │Dur. = 6│  │Dur. = 2│
└────────┘    │ Cycle Time = 2│        └────────┘  └────────┘  └────────┘
              └──────────────┘              ↓           ↓           ↓
┌────────┐    ┌──────────────┐         ┌────────┐  ┌────────┐  ┌────────┐
│ Act. 2 │────│ BOW Act. 2   │         │Task 2.1│→ │Task 2.2│→ │Task 2.3│
│        │    │ Target = 5   │         │Dur. = 3│  │Dur. = 9│  │Dur. = 3│
└────────┘    │ Cycle Time = 3│        └────────┘  └────────┘  └────────┘
              └──────────────┘
```

<u>Actions and their global bills of work</u>         <u>Associated Tasks</u>

**Figure 8.9** Infinite loading overlap transfer task and constraint graph

account the batch size, and the estimated cycle time and estimated performance parameters of the assigned resource(s). If the batch size is one or two, then the same amount of tasks is created, respectively. Each of the tasks represents the time taken for one part to be processed by the estimated assigned resources. If the batch size is equal to or greater than three, then three tasks are created. The first and last tasks correspond to the time for the processing of a single product, and the middle task, to the processing time of the resources multiplied by the batch size minus two (representing the time to process the remaining parts, see Figure 8.9).

The scheduling method is applied to the workshop or cell(s) to schedule the assigned bills of work, and is called global scheduling.

The tasks that were created before are then linked as follows. If the transfer method is batch transfer, then the tasks are first linked according to their logical order in the process graph. Then the tasks are linked to one another, according to the order in which they were scheduled.

If the transfer method is overlap transfer, the first task of each bill of work is linked to the first task of the following action. Likewise, the last task is linked to the last task of the following action. Furthermore, the tasks of a bill of work are linked to one another, to represent the sequence of the estimated resources. Then the tasks are linked to one another, according to the order in which they were scheduled.

Once the task and constraint graph has been constructed, the earliest and latest starting and finishing dates of the tasks are computed. Then, the loads of the workshop or cell(s) are computed.

# The Scheduling Module

This information is then shown to the user, who can then modify the graph at will, typically by permuting the order of the bills of work of the resources, changing the dates to see their effect, or splitting or grouping. The user is also able to perform what-if analyses at this level.

**(iii) Finite loading**

Finite loading was provided for as a precise schedule computing aid, both at the workshop, and the resource levels. Both batch and overlap transfer infinite loading were provided for. It allows the determination of the estimated completion dates of the production orders for the specified period.

Finite loading is equally computed for a single cell level at a time; therefore, if the user wants to compute the loads of one of the subcells of the workshop, he must first place himself at the level of the subcell.

The computation of the dates for batch transfer finite loading is equally done through the construction of a task and constraint graph of the cell.

The method is essentially the same as for infinite loading, except that here, as opposed to infinite loading, the resources assigned to the actions of the process graph are scanned, and not just the actions themselves. Then the bills of work of the assigned resources (local bills of work) are scanned. The tasks are then constructed from these local bills of work (see Figure 8.10).

The scheduling method is then applied directly to the concerned resources, as opposed to the workshop or cell(s), which don't necessarily have the same scheduling requirements. The graph is then linked, as for infinite loading, except that the

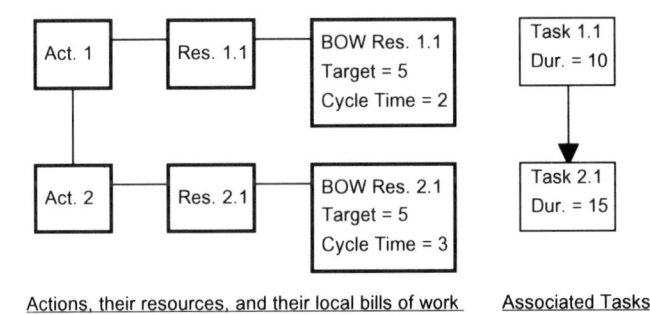

| Act. 1 — Res. 1.1 — BOW Res. 1.1 / Target = 5 / Cycle Time = 2 | Task 1.1 / Dur. = 10 |
| Act. 2 — Res. 2.1 — BOW Res. 2.1 / Target = 5 / Cycle Time = 3 | Task 2.1 / Dur. = 15 |
| Actions, their resources, and their local bills of work | Associated Tasks |

**Figure 8.10** Finite loading batch transfer task and constraint graph

scheduling constraints are those of the assigned resources. Then, the dates and loads are computed, as for infinite loading, with the habitual interventions from the user.

**(iv) Scheduling heuristics**

The scheduling module presented in this book is object-oriented. This means that an independent local strategy is available for each resource, if required. The strategies presented here have been chosen with this in mind.

It should be noted that the application of a strategy will depend on a variety of factors, in particular the general structure of the workshop, and production management working methods and strategies.

If one takes, for example, a job-shop scenario, with one or several production bottlenecks (a typical OPT-like scenario (optimized production technology, developed by Dr E. Goldratt (Goldratt and Cox, 1984)), one would logically only schedule the bottleneck resources. The other resources would simply be left to operate in the FIFO mode.

This also holds true for a classical permutation flow-shop automated assembly line, where the bottleneck resource determines the scheduling strategy.

If required, a number of resources can be scheduled separately, to allow the optimisation of a number of different criteria at the same time. Let's return to the OPT-like scenario; the critical resource (a bottleneck in this case) has already been scheduled. Even though the delivery dates are the dominant criteria here, one or a number of resources are very expensive to operate continuously. In this case, these resources would equally be scheduled in accordance with this criterion.

To sum things up, the scheduling done on the resources can be considered as being either active or passive. If active, the chosen strategy for the resource is applied to the list of bills of work assigned to the resource. If passive, the strategy will be to operate in the FIFO mode, or in other words, not to change the list of pending bills of work, which have been assigned in the right order, the scheduling being already complete at the level above with global scheduling.

In this manner the flexible architecture of the module allows an easy, modular and efficient scheduling of the workshop.

One of the major constraints for the module is that it is implemented in an industrial computer, or supervisor. As the supervisor has numerous other real-time functions in use at the same time, it has only limited computational power available for scheduling at any one time. To be compatible with the real-time aspirations of the module, the strategies must be cheap on computational time.

The chosen strategies, listed here below, take the above-mentioned constraints into consideration:

- *FIFO (first-in–first-out)*. This scheduling strategy simply entails the acceptance of the list of bills of work in the order in which they are down-loaded.
- *LIFO (last-in–first-out)*. This scheduling strategy entails the inverting of the order of the list of bills of work with regards to the order in which they are down-loaded.
- *SOT (shortest operation time)*. This scheduling method entails the sorting of the orders according to their operation time, from the shortest to the longest.
- *LWKR (least work remaining)*. This scheduling method is a generalisation of the SOT method, which considers the operation times of the order's following resources.
- *Due date*. This scheduling method entails the sorting of the orders according to their due date, from the earliest to the latest.
- *Starting date*. This scheduling method entails the sorting of the orders according to their starting date, from the earliest to the latest. The job with the closest starting date is launched first, thus avoiding delays.
- *STR (slack time remaining)*. This scheduling method entails the sorting of the orders according to their slack time (time between the earliest and latest starting dates), from the lowest to the highest. The job with the smallest slack time is launched first, thus avoiding delays.
- *STR / Op (slack time remaining per operation)*. This scheduling method entails the sorting of the orders according to their slack time (time between the earliest and latest starting dates), divided by the number of operations to be done; from the lowest to the highest. The job with the smallest slack time per operation is launched first, thus avoiding delays.
- *CR (critical ratio)*. The critical ratio is the ratio between the difference between the due date and the current date, and the sum of the times of the operations still to be done. The jobs with the smallest ratio are processed first. Those that have a ratio smaller than one are already late.
- *QR (queue ratio)*. The queue ratio is the slack time remaining,

divided by the expected waiting time. The jobs with the smallest queue ratios are treated first.
- *WINQ (work in next queue).* In this method, the following operation, whose resource has the smallest list of pending orders (by pending orders, one means the sum of the targets of the bills of work), will be treated first. This method tends to reduce the through time of the orders, and decrease the unproductive times of the resources.

Another possibility is to implement a user-defined heuristic on a separate computer, and adapt it so as to provide the best current schedule during the search of the solution space of the problem. In this manner, if the user needs to start producing straight away, he or she can take a rough schedule that is quickly computed by the method, download the first bill(s) of work, and leave the scheduler to keep optimising the rough schedule. This can also be made easier by adding the corresponding constraints, thus reducing the feasible solution space. Later on, the selected schedule can be modified to take into account the refinements obtained in the mean time. This is particularly feasible if the number of surprises at the shop floor is quite small. A good example of a scheduling method particularly adapted to this approach is TABOU (Widmer, 1994).

If, however, the number of surprises is considerable, this approach is no longer the most appropriate (see Castelain *et al.*, 1994) for an automobile final assembly line example of a real-time industrial scheduling example with numerous surprises and constraints. In this case, simpler, faster heuristics that always give a schedule are more appropriate, such as the ones presented above barring a few detail modifications, or conversely, special application-dependent methods have to be developed.

To sum things up, the above-listed strategies chosen during the project are sufficient for the needs of current industrial users in general, and can be easily added to, if required.

## 8.5 Conclusions

The scheduling module presented in this book not only replies to industrial needs in general, but has applied them as an important guideline during its elaboration.

These needs are nevertheless changing at a significant rate, due to changing markets and strategies and, as importantly, to new system supervision hardware and approaches. This module already satisfies most of these needs. However, if the size of the scenario is very big, the module would greatly profit from increased distributed architecture capabilities. These architectures are supported by new and innovative scheduling approaches, particularly in the field of artificial intelligence (Butler and Ohtsubo, 1992; Hadavi *et al.*, 1992, Roubellat *et al.*, 1994).

A distributed architecture has numerous advantages above a traditional one, particularly with regards to its modularity and ease of representation of constraints. With such an architecture, the strategies and constraints change with each level, as they naturally do in a real production environment. The closer one gets to the production management side, the more global optimisation and organisational constraints are important. The closer one gets to the production side, the more the technological and manufacturing constraints become important. These different constraints can be represented in a sufficiently detailed manner with a distributed architecture, and, even more importantly, be more easily understood. This leads to better system representation and scheduling performance.

In its present form, the module has most of the basic requirements to be compatible with such an architecture. What it still requires is the definition of a higher level communication protocol between the different local scheduling modules situated on a factory floor, to further automate scheduling. This is relatively simple if one applies a top-down approach to the scheduling and the transmission of the constraints and parameters. In the other direction, numerous feasible strategies can be applied, such as requests for modifications to the higher-level schedules due to unforeseen constraints or events. Furthermore, the problem of information integration and filtering has to be resolved, to render the information flow more efficient. A rather significant result of our concurrent engineering approach is that the architecture and databases that were defined in this book are already adapted to this approach.

A more generalized application of the architecture established in this book, together with the new above-mentioned protocol, via a more global use of the inherent hierarchy, would result in increased scheduling possibilities. With the provision of these new functionalities, the scheduling approach of this module could be applied to systems of far greater complexity. Nevertheless, the module already replies to most current industrial needs, which was its main requirement.

A further evaluation of the capabilities of this architecture would be most beneficial as a direction of research and define a tool for big, complex scheduling environments.

## 8.6 References

Anon (1994) Machine Design Staff Report, Manufacturing planning goes world class, *Machine Design*, April.

Butler, H. J., Ohtsubo, H. (1992) ADDYMS: architecture for distributed dynamic manufacturing scheduling, in *Artificial Intelligence Applications in Manufacturing*, AAAI Press.

Castelain, E., Hammadi, S., Ohl, H., Gentina, J.-C., Riat, J.-C. and Yvars, P.-A. (1994) Ordonnancement réactif: Génération de la séquence de véhicules en entrée de ligne de montage automobile, *Journées d'étude: Ordonnancement et entreprise: Applications concrètes et outils pour le futur*, 14–15 June, Automatic Research Group CNRS.

Goldratt, E. and Cox, J. (1984) *The Goal: Excellence in Manufacturing*, North River Press, New York.

Hadavi, K., Hsu, W.-L., Chen, T. and Lee C.-N. (1992) An architecture for real time distributed scheduling, in *Artificial Intelligence Applications in Manufacturing*, AAAI Press.

Lauvigne, O. (1994) Automates Programmables: Se Diviser pour Mieux Superviser, *Usine Nouvelle*, No. 2453, 21 April.

Roubellat, F., Billaut, J.-C., Villaumie, M. (1994) Ordonnancement d'un atelier en temps réel: D'Orabaid à Ordo, *Journées d'étude: Ordonnancement et entreprise: Applications concrètes et outils pour le futur*, 14–15 June, Automatic Research Group CNRS.

Saunders, D., Crockett, M., Novels, M. and Hackwell, G. (1990) Generation of material flow conrol via simulation, *Second Internl Conf. Factory 2001—Integrating information and material flow*, 10–12 July.

Taylor, A. (1990) How intelligent KBS, expert, real-time systems will provide the key to integrating information and material flow of Factory 2001, *Second Intl Conf. Factory 2001—Integrating information and material flow*, 10–12 July.

Van Dam, P., Gaalman, G. and Sierksma, G. (1993) Scheduling of packaging lines in the process industry: An empirical investigation, *Intl J. Production Economics*, **30–31**, 579–589.

Verdebout E. (1992) *Pilotage algorithmique du changement de campagne de production dans une installation flexible d'assemblage automatique*, Thesis No 1013 presented at the Dept. of Microengineering, EPFL (Swiss federal Institute of technology, Lausanne).

Widmer, M. (1994) *Modèles Mathématiques pour une gestion efficace des ateliers flexibles*, Thesis EPF-Lausanne, Presses polytechniques et universitaires romandes.

# 9 *The Flow Control Module*

## 9.1 *Introduction*

The industrial scenarios, on which we have based the specifications of the CAD method for industrial assembly, manage production by means of batches; even though batch size is ever decreasing, just-in-time techniques are an exception in which a lot size of 1 is applied.

Production orders are issued for a batch of items rather than separate items:

- Management of families of product variants can be achieved if one is able to manage changes of production runs.
- Flows of identical parts are successively processed by the resources. Within the resources, the flows are possibly combined to produce other new flows: this is the assembly action itself.

The flow control module is closely related to the scheduling module:

- It performs the order release according to the chosen schedule.
- It also processes the real-time decisions that enable the flows of parts to move within the layout.

This module concerns two kinds of elements of the layout model:

- The resources:
  — By means of the list of bill of work entities associated to them, the flow control module will manage the resources locally and give them a limited autonomy of decision.
  — By means of the routing rules possibly associated with

them, the module will also enable the resources to prompt the supervisor to return the chosen destination of the parts that leave the resources towards other elements.
- The switching elements:
  — By means of an input switching rule associated with them, the module will enable the switching elements to decide which part may enter them, that is, to manage the priorities at their inputs.
  — By means of a list of output switching rules, the flow control module will also allow these elements to decide to which of the next elements a part will be given, according to a series of criteria.
  — Among these criteria, a switching element may request a decision from the supervisor if the locally available information is not sufficient to process a pertinent decision.

Figure 9.1 displays the hierarchy of entities concerned by the flow control module.

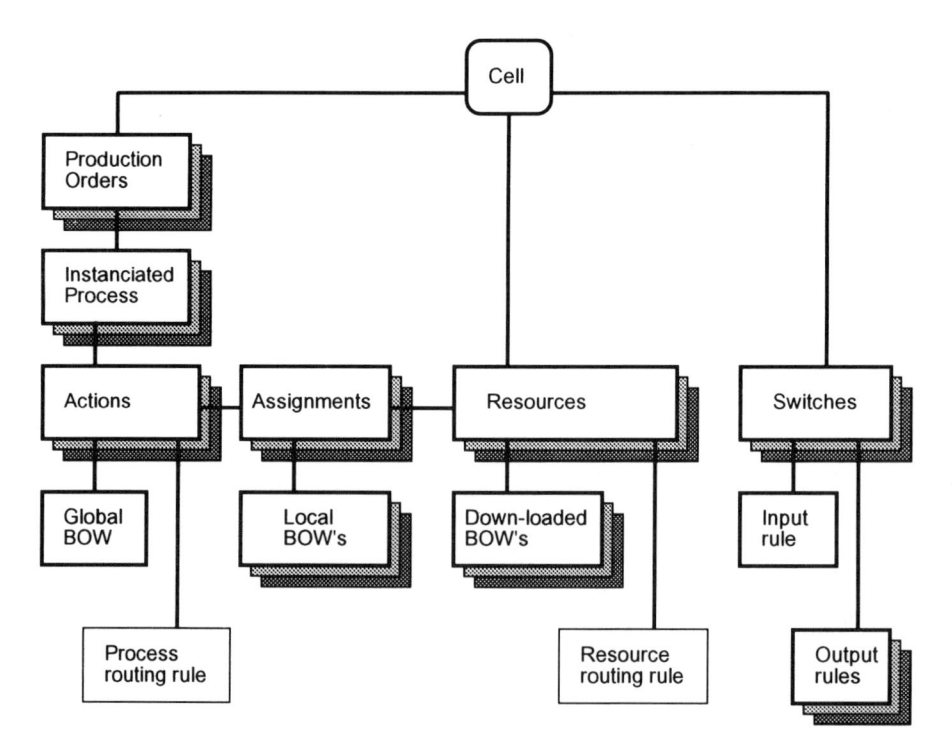

**Figure 9.1** Hierarchy of the entities concerned by the flow control module

# The Flow-control Module

Two kinds of decision making are performed by the module:

- Local decisions are taken directly by the resources whenever the locally available data is sufficient to do so.
- Global decisions are taken by the supervisor in any other case.

Finally, the module will be used in two distinct situations:

- During layout design, it is used with the simulation module. The flow control parameters will be completed as the design progresses.
- The same module is used during the production phase. The parameters resulting from the off-line study are transmitted from the simulation platform to the real layout once its controllers are ready for use.

By using a concurrent engineering approach, the design phase of flow control may start very early with an incomplete and provisory layout.

This allows a gradual design of the control system concurrently with the design of the product and resources, and a feedback of the flow control point of view to the other aspects of the design.

The flow control module belongs to both the off-line and on-line parts of the CAD method; its connections to other modules are shown in Figure 9.2.

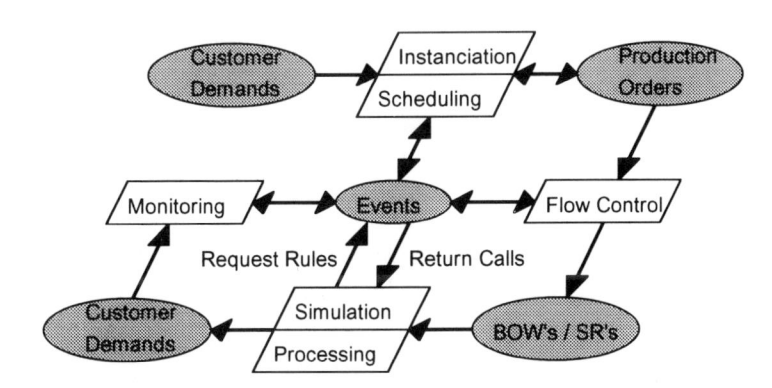

**Figure 9.2** Situation of the module within the CAD method for industrial assembly architecture (BOW: Bill of Work; SR: switching rule)

The module associates data with the elements of the layout database. It receives orders and bills of work instantiated by the scheduling module:

- The production orders may be issued by a scheduler, a production planning system, or edited by the user.
- The bills of work may be created by the scheduler when assigning instantiated actions to each of the chosen resources.
- They may also be resident within the controllers of the resources if one doesn't intend to use the processing capabilities of the module.

The switching rules define which real-time decisions concerning the traffic and routing of the objects are to be processed within the controllers of the switching elements during production.

The flow control decisions can also be global within the supervisor; in this case, requests for decisions are posted by means of events, the answer being transmitted from the supervisor down to the local controllers.

Finally, the flow control module receives events from the scheduling module when the schedule is modified, and from the layout if important changes of element state occur. The flow control module will itself post events to the scheduling module when system errors occur, for example jamming of the resources.

## 9.2 Users' Needs

The questionnaires filled in by industrial companies have clarified the scenarios of production to be foreseen by the flow control module:

- Manual assembly of small series is not foreseen.
- Manual or semi-automated assembly of medium series of product families:
  - cycle time from 0.5 to 1 min;
  - possibly assisted manual assembly requiring lower skill;
  - possible just-in-time management requiring no scheduling at all;
  - management of the flows of parts in a flow shop;
  - numerous variants are to be taken into account.
- Mass production assembled automatically with at least several variants:

— cycle time from 0.5 to 8 s;
— fully automated assembly;
— MRP-like scheduling with batch sizes from 1000 to 1000 000.

The users' requirements can be summarised in the following categories:

- Minimise the necessary control and communication hardware:
  — comply with the architecture of our general CAD method;
  — consider the supervisor (industrial computer that controls the assembly line) as being the global cell controller;
  — during simulation, consider a simulated supervisor;
  — consider the possible lack of a scheduler.
- Handle production constraints and expected trends:
  — allow for safe human intervention;
  — manage families of product variants;
  — connect the flow control module to the scheduling module;
  — increase flexibility;
  — provide strategies that decrease loss of components.
- Re-use existing methods:
  — provide a support for user-defined strategies;
  — allow routing strategies (details hereafter).
- Enhance current flow control functionalities:
  — reduce human intervention during the set-up of a new production run;
  — provide enhanced error recovery capabilities;
  — display the on-going processes (now, only the resource state is displayed);
  — allow for the changing of the resources assigned to a given operation;
  — all the users mentioned the need for rescheduling to reply to problems.

## 9.3 State of the Art

Flow control is understood by most industrial companies as the management of production orders in the process of manufacturing and assembly.

In automated assembly lines, flow control is trivial if the parts (or the pallets or trays which carry them) are not mixed in the layout. In this case, the identification of a part is directly related to the location it is at, and to the current production status within the sequence of orders.

However, if the flows of parts are mixed, it becomes necessary to identify them and to direct them towards their proper destination.

Identification systems on pallets, keeping a trace of the assembly sequence, are used in many cases. The labels can be static or dynamic:

- Using a static label will restrict the flexibility, or oblige the user to define interrogations to a centralised system in a dynamic manner.
- However, the dynamic label systems generate a higher system cost which must be taken into account, especially when dealing with a large amount of pallets.

In many cases, assembly systems based on a distribution by means of conveyors and switches will locate some intelligence at the switches.

Local algorithms will discriminate the pallets destined for a given machine or station, according to a destination identifier, a state of progress, or any other criteria.

The management of the change of production run is an important flow control feature, since human intervention within the resources of the system is to be limited. Nowadays, this management can be achieved in several possible manners according to its frequency:

- the easiest manner is to clear the whole system of any part, so that the two successive productions are never mixed—this method is reserved for large series.
- in the flow-shop workshop product flow organisation, which is frequent in automated assembly, two successive productions will follow the same assembly process, except for the variants of components and operations.
- The change of run will involve a clear separation of the successive flows, for example by means of dummy parts of a very bright colour.
- This manual management gives trouble with today's decreasing lot sizes.
- in semi-automated scenarios, the need for new solutions is clearly expressed by industrial companies we worked with and is the purpose of this section.
- in most flexible scenarios, for example in the final assembly of cars, very large computations are achieved in real time but such a complexity surpasses the intention of this section.

# The Flow-control Module

## 9.4 Functionalities and Methodology

### 9.4.1 Introduction

The functionalities of the flow control module can be divided into the following complementary categories:

- functionalities related to production, which manage the bills of work;
- functionalities related to traffic decisions;
- functionalities related to routing decisions;
- functionalities related to human intervention.

Each of these categories will be briefly recalled hereafter.

Figure 9.3 illustrates the difference between traffic and routing:

- traffic is a local problem;
- routing is a global problem.

- Traffic is a local problem,
- Routing is a global problem.

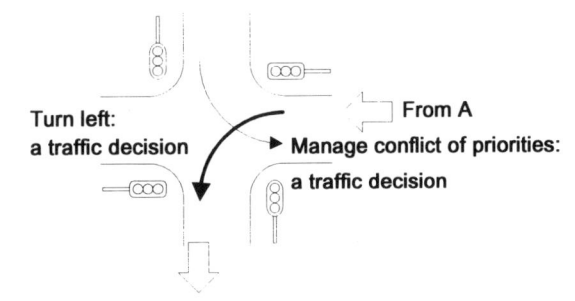

Turn left: a traffic decision

From A

Manage conflict of priorities: a traffic decision

To B: a routing decision

**Figure 9.3** A crossroads to illustrate traffic and routing

### 9.4.2 Functionalities Related to Production

From the global point of view, the corresponding functionalities are:

- Communicate the bills of work to the assigned resources when:
  —these bills of work have been sent by the scheduling module,
  —the corresponding resources are ready to receive them.

- Possibly manage the compensation of defective parts to avoid losing parts or machine cycles through excessive production, and reduce the minimal batch size that can be processed economically.
- Monitor the production in order to detect changes of production run and process the corresponding changes of data.

From the local point of view, two kinds of functionalities will define the decentralised part of the flow control module:

- Receive the updated parameters, reply with updated variables.
- Maintain an up-to-date local SPC (statistical process control; supervision of the performance of the resources of the workshop) and SQC (statistical quality control; supervision of the production of the products).

The resources should avoid sending superfluous interrupts to the supervisor, which would intolerably slow down the process due to response delay.

Local management of SPC/SQC data (state integrators, part counters) and of the routing strategy will provide a limited autonomy to the resources. As a result, the stations may be developed with no need for the supervisor: they can be disconnected from the system. In the same way, a temporary interruption of the supervisor doesn't immediately stop the system.

## 9.4.3 Functionalities Related to Traffic Decisions

Some features of the module are related to the *traffic* function of the switching elements and devoted to the supervisor:

- Define the parameters that permit the automated processing of traffic decisions in a generalised manner.
- Establish a communication link between the supervisor and the switching elements, in order to monitor the local traffic decisions when required.
- Establish another link between the supervisor and the switching elements by means of interrupts, if data processing is required from the supervisor.
- Give an access for user-defined global traffic strategies to be prompted by means of interrupts.

# The Flow-control Module

Some other features are devoted to the switching elements and are thus local:

- Establish the priority between the objects (parts and pallets) when a conflict arises at the inputs of a switching element.
- Choose the output channel for a given object when it leaves a switching element.
- Give an access for user-defined local traffic strategies and send requests for global decisions when required.

## 9.4.4 Functionalities Related to Routing Decisions

When a mobile object leaves an element of the layout, it is sent to a set of other elements that can be determined by an analysis of the layout configuration; this is very much a graph-scanning related problem.

The purpose of the routing decisions is to choose a destination that defines a subset of this set, for each mobile object (part or pallet) leaving a resource. The choice depends upon the assembly process (next action to perform), and the subset of resources that have been assigned to the next action for the production order this object belongs to.

A destination information can be associated, if required, with a mobile object. This can be realised in a dynamic or static manner, with or without centralisation within the supervisor: the decisions will remain the same in any case. The corresponding global functionalities are:

- Define the parameters that permit the automated processing of routing decisions in a generalised manner.
- Establish a communication link between the supervisor and the resources, in order to monitor the local routing decisions when required.
- Establish another link between the supervisor and the resources by means of interrupts, if data processing is required from the supervisor.

For the same purpose, the local functionalities are the following ones:

- Establish the destination of each part or pallet that leaves the resources when required.
- Give an access for user-defined local traffic strategies and send requests for global decisions when required.

### 9.4.5 Functionalities Related to Human Intervention

Two kinds of users may interact with the flow control module:

- highly skilled persons who configure the system at the local and global level, and solve problems during production;
- operators who interact when fixing the jams and during changes of production run; they are naturally in charge of the manual stations.

An explicit industrial need is to make human intervention safe by limiting the access to any features by means of passwords, access keys, ... In addition, the purpose of the present section is to define the kind of access that can be useful when using the flow control module:

- Modify locally and globally the data during configuration.
- Provide a centralised access to the local data.
- Allow for temporary modifications of the flow control data that enable user intervention within the resources.

This last functionality is worth paying some more attention to.

In order to provide safety by means of automated control, the intervention of the user must always be permitted. Nevertheless, any human intervention should be made safe. This is the reason why an analogy can be found between the flow-control module and railway signalling hardware. Railway signalling always allows the addition of supplementary restrictions by the user but never permits an easy suppression of an automated restriction.

### 9.4.6 Methodology

**(i) Approach within the CAD method for industrial assembly**

**Define the same functions off-line and on-line** The flow control module will be used in two distinct situations: the virtual world of the simulated environment, and the real world of the actual production line.

Our approach assumes that the results of simulation are transmitted to the supervisor: flow control must be separated from any simulation feature.

When realising the off-line and on-line modules, the programmer will face different programming constraints:

- Off-line programming of the module will create a close interaction between the simulated model and the module by means of events or function calls.
- On-line programming of the module will lead to interventions within both the cell supervisor and the controllers of the resources. Within the PLC's, wide use is made of sequential programming techniques such as the Grafcet language. Research at the EPFL (Swiss Federal Institute of Technology, Lausanne) has proved by means of a prototype that such functions can be realised within a PLC.

These programming constraints will lead to distinct programs when the module is used off- and on-line, but the functions and models will be identical. Communication between these two modules is to be achieved between the CAD station and the supervisor. In fact, only the models and data will be transmitted in both directions.

**Automation and safety: the railway analogy** We had to face two opposite trends when defining the solution to the expected functionalities:

- Increase the automation of the flow control module (traffic, routing and production functionalities) due to production constraints and the need for increasing complexity.
- Increase safety and interactions of the user with the module and the system.

These two distinct and contradictory goals can be found in exactly comparable terms in another activity, railway signalling:

- simplicity and reliability of the hardware even though the automation level is ever increasing;
- creation of safety by means of the automation of any repetitive task while allowing for human intervention at any moment.

The analogy of flow-control and railway signalling has been explored in detail in previous research at the EPFL.

The main result to be transposed in the assembly environment is the "sectional release route locking", explained in more detail later on.

This signalling technique of the railway environment was

developed to achieve the same goals: increase the flexibility given by the points and crossings of the railway network while ensuring the absolute safety given by the block system.

**Standard modules customised by parameters** Another goal was aimed at: the reduction of the programming effort through the definition of standard procedures.

The elements of the layout model have been identified as objects. In the same manner, the flow control behaviour has been standardised in the two main categories of elements:

- the resources (stations, feeders, exits, etc.) in which production takes place through the modification of the flows of parts and pallets;
- the switching elements in which priorities have to be respected and in which the parts and pallets are to be directed towards other resources.

Both categories will receive a standardised flow control treatment. This means that its development can take place:

- either from a very early stage of the project, thus enhancing the concurrency of the system and product design,
- or, be considered as a standard software product of the supplier company, thus meaning that this design phase is simply suppressed.

This will cut down the development time of such an installation.

These standardised functions will receive their actual parameters at the last moment during the project, thus permitting their optimisation in the simulation environment before the final downloading.

**Decentralising the intelligence** Hardware costs are ever decreasing, compared to their capabilities; strong efforts are made in many projects to prove that a decentralised approach is valuable and cost-effective. Nevertheless, a compromise must be found when considering the existing industrial situation and the widespread use of low-cost PLCs and networks with limited capabilities.

Based on EPFL's research work on minimal flow control functions to be associated with the local controllers, this module of the CAD approach intends to:

# The Flow-control Module

- give a local autonomy to the resources in order to permit production with no need to interrupt the supervisor, thus allowing for the development of a station away from the site;
- stop the production by means of de-activation of the controller as soon as any information is required from the supervisor, thus permitting user-defined strategies based on its interrogation through synchronous interrupts;
- manage automatically the change of production run when the resources are technologically prepared for it.
- This goal will be achieved by taking advantage of the railway analogy exposed above.

These points will lead to the solutions we present in the next section.

## (ii) New concepts of flow control

**The railway analogy** The analogy between flow-control in assembly and railway signalling, introduced in previous research work at the EPFL, is an innovative concept introduced in the CAD method discussed in this book (Figure 9.4).

Naturally, this analogy has some *limitations* which must be foreseen:

- The "trains" may get mixed up in the assembly system!
  Parts are sent from a source to a destination independently.
  Instead of taking the decisions once per train within the switching elements, such decisions must be taken for each mobile object.
- There may be more than one source or destination element!
  Resources can be installed in parallel in order to achieve equivalent tasks.
  Instead of being a sequence of locations, the path between

| Railway signalling | Assembly workshop |
| --- | --- |
| Scheduled circulation | Flow |
| Vehicle | Part or pallet |
| Departure station | Source: set of resources |
| Arrival station | Destination: set of resources |
| Composition of a train | Assembly within the resources |

**Figure 9.4** Analogy between railway signalling and flow control in assembly

the source and destination has possibly more than one root and/or more than one leaf.
- There may be cycles and circuits within the paths!

    Cycles are encountered when stations are shunt connected on a transfer line.

    Circuits are encountered when parts can be stored on a conveyor loop.

    Particular attention will be required when defining the flow-control methods, so that the changes of production run take into account the mobile objects that are still travelling somewhere in the transfer loop.

Even though these limitations must be taken into account, it is interesting to take advantage of the railway methods when controlling an assembly installation:

- The concept of a *block system* applies to the resources.

    The detection of the flows can be achieved as in certain Swiss mountain railways, by means of occupation counters dedicated to each particular flow. These counters will enable the comparison of the number of processed parts with the production targets. Once the target is reached, the production run is finished.

    Nevertheless, quality problems must be taken into account: parts can be rejected, directed towards a repair station or returned from such a station.

    Compensation strategies will take these surprises into account.
- The concept of block-system *signals* associated with assembly tasks enables the control of set-up management.

    In our model, these signals are dedicated to a given flow. When a batch has been processed, turning the signal of the next batch to "stop" will enable the set-up and prevent the mixing of flows.
- The concept of *sectional release route locking* is also applicable. In railway networks, this consists of assigning a path within the network for a given circulation, thus forbidding its access to any other train. Once the occupation has been detected and then freed, each section of the path is successively released to allow for its occupation by another train that would otherwise be incompatible with this one. The assignment of a path for the next train is permitted at the earliest possible date. In the workshop, sectional release route locking

is interesting when the flows cannot be mixed up, for example when machines need a set-up. It enables the management of an amount of batches that surpasses the memory capacity of a limited PLC: the bills of work can be stored in advance, then downloaded when required.

To achieve sectional release route locking in assembly systems, a sequence of successive locking states is implemented in the bills of work:

1. Store the bill of work in the supervisor as long as the controller of the resource is not ready to receive it.
2. Download the bill of work with a STOP signal to prevent parts from leaving the resource, as long as the next section is not ready.
3. Open the signal once the next section is ready.
   Possibly, manage an intermediate starting phase if the resource does not need to be emptied between two successive runs.
4. Count the parts that leave the resource during the production phase, until this amount equals the production target.
   During this phase, exchange events with the supervisor to compensate for defective parts and possibly decrease the production target due to the production of parallel resources. This data exchange does not need to be synchronous with the production cycles: this is the reason why it has been called *asynchronous polling* of the resources by the supervisor.
5. Finish the production phase and prepare for the next set-up or intermediate phase if no emptying phase is required.
6. Once requested by the supervisor, delete the terminated bill of work and shift the other bills of work ahead one place to prepare a new reception.

**The switching rules** The switching rules represent the decisions that can be processed by a switching element if it considers only the data that is locally available:

- the information associated with objects that are currently waiting at its inputs,
- the existence of available places at its outputs,
- the data associated with the object it may contain.

Such decisions are to be separated into two categories:

- input rules, which represent the choice of the next mobile object to take within the switching element,
- output rules, which represent the choice of the direction towards which the currently contained mobile object is to be sent.

The innovative concept introduced by our approach is to consider such decisions as *parametered objects* applicable within PLCs. The advantage is that a program can be prepared long before the layout is even known, as demonstrated in previous research at the EPFL.

The input switching rule is driven by *conflicts*: its purpose is to define a set of decreasing priorities among the input channels (or links) of the switch. A signal is associated with each channel and can be controlled by the user: this allows human intervention.

At the outputs, the problem is a bit more complicated since several decision criteria can be involved:

- flow rules that apply to all the parts and pallets of a given flow;
- destination rules that apply to all the parts and pallets destined to a given set of resources;
- user-defined rules that may apply to any pertinent information required by the user and locally available, thus providing an available access to include his or her experience;
- finally, one default rule that applies when none of the above rules applies to a given contained part or pallet.

This rule will help the user of off-line simulation to start during the early stages of flow control definition, even though he or she doesn't know the layout precisely.

When several kinds of rules are defined together, a priority must be defined between them; the more precisely the mobile object fits a given criterion, the higher the priority of the rule. So far, we have defined the way a switching element will act to choose a rule when it knows the information associated with the object. The rule itself will be defined by its algorithm of choice for the output channel among several basic algorithms:

- Decreasing priority rules will define a subset of the available outputs, sorted by decreasing priorities.
- Pondered cyclic rules will define a sequence of pairs (channel, weight) and try to respect it.

# The Flow-control Module

- Part state rules will define a sequence of pairs (channel, state) and are especially interesting when discriminating the pallets to be reworked.
- Request rules that post a given request to the supervisor, which triggers the execution of a global rule that responds with the channel to be used.

Since it is based on interrupts to the supervisor, this last kind of rule applies only if the capabilities of the hardware enable its processing without any intolerable delay that would slow down the whole process. The modularity of the rules and the access to the traffic signals by the user make such rules a very powerful tool to control the flows in any kind of scenario.

Let's take the example of the shunt connected stations (see Figure 9.5.).

- The model of the derivation is shown in Figure 9.6.
- The default output strategy of the input switch B is to proceed straight, but if the objects are destined to the station, they are directed towards it if there is enough room.
- The input strategy of the output switch A is to prefer pallets coming from the station to those coming from the pallet circuit in case of conflict: this accelerates the output of the station.

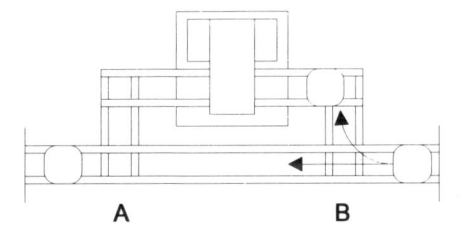

**Figure 9.5** Diagram of a shunt connected station

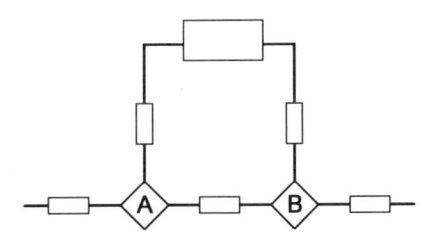

**Figure 9.6** Its model

| Rule | Element | Kind of rule | Algorithm | Remark |
|---|---|---|---|---|
| 1 | A | Default input rule | Prefer parts from station | |
| 2 | A | Default output rule | Give parts to output | Trivial |
| 3 | B | Default input rule | Prefer parts from input | Trivial |
| 4 | B | Default output rule | Parts proceed straight | |
| 5 | B | Destination output rule | Parts destined to the station enter the station if possible, or proceed straight otherwise | |

**Figure 9.7** Minimal set of rules for this example

- Five switching rules, including two trivial rules are sufficient to manage the case (see Figure 9.7).

**The routing rules** The routing rules are an innovative solution to manage the choice of the destination of a mobile object when it leaves a resource: it is also based on previous research at the EPFL.

Let's recall that a destination usually identifies one of the inputs of a given resource. The problem of routing consists in deciding upon the next destination that should be associated with the departing mobile objects. Such a decision is to be processed each time a mobile object leaves a resource: this decision must be processed synchronously with production. In most cases, the information of the destination associated with the flows is sufficient to process the decision locally.

When a bill of work is transmitted to a local controller, it includes a set of resulting flow entities, each of them including an individual destination that applies to the mobile objects of this flow that leave the resource. This general rule applies to most of the production scenarios.

It should be noted that a destination does not always identify a single resource input: it may also represent a set of resource inputs, thus applying also when several resources are to work in parallel. If the user wants to define his or her own routing rules, he or she will have to define another rule. For example, if identical objects are to be sent to a variable destination according to other criteria, a request rule may be user-defined. In this case, the request code will trigger the corresponding algorithm within the supervisor and return the chosen destination in real time. In this case, the whole production system may be slowed down by this interrupt.

As illustrated in Figure 9.1, the global routing rules implemented in the supervisor can apply either to a given resource or to a given action of the process.

**Managing the request rules** The request rules are those for which the information available at the local level is insufficient to process the decision. Such a management is to be processed differently off-line and on-line.

In the off-line simulation environment, the supervisor intelligence is associated with the cell aggregation.

Real-time decisions of the cell concerning a given element can be directly called within the simulation package: the returned result will be directly obtained within the format of the function call. However, when defining the same functions in an on-line environment, the cell supervisor and the PLCs are physically separated. Such a call between two different controllers is implementation-dependent. Two main ideas can be foreseen to realise it:

1. Generate an interrupt of the supervisor from the PLC including the request code, which triggers the corresponding reaction of the supervisor and returns the expected result. This may create a large amount of interrupts within the supervisor, thus obliging it to manage a large task to manage the interrupts.
2. Create a mailbox of several variables exchanged between the PLC and the supervisor and monitor the change of the variables. This allows the writing of the request code in a reserved register of the PLC, which in turn triggers the reaction of the supervisor which writes the answer in the corresponding reserved register. The modification of this result register allows the cancellation of the request.

**The compensation strategies** The compensation strategies are involved when defective parts are produced. When such a part is rejected by the inspection stations, the number of parts is modified: the loss should be compensated. The compensation strategies usually belong to the following categories:

1. *just-in-time* scenarios in which single demands generate single products, which fits with relatively slow assembly processes but generates lots of management in fast assembly systems;

2. *preventive* compensation that takes into account the statistical risk of defect but never insures the exact amount of finished products, which fits with small families of variants but generates intolerable delays with large numbers of variants;
3. *curative* compensation orders added to the current schedule with the amount of effectively rejected parts.

These three strategies can be used with others to define a catalogue of compensation strategies. These informations are not exchanged between the different modules.

The EPFL has defined a new strategy based on two phases, which complies with the needs of medium size production:

- From the beginning of the production order, each occurrence of a defect generates an incrementation of a dedicated counter within the resource that created it. This can be the rejection counter or the bad part counter, depending on whether an ejection is used or not.

  When the defect is detected by the supervisor, a compensation feedback is triggered: each action of the process is requested to increment the production target. This is achieved by means of a tree-like propagation within the instantiated assembly process, illustrated in Figure 9.8. This propagation stops if a resource has already started to produce something else.

- At the end of the production run, such a compensation is impossible. The user may then use one of the usual techniques (prevention by adding a given amount of extra parts, or a compensation order).

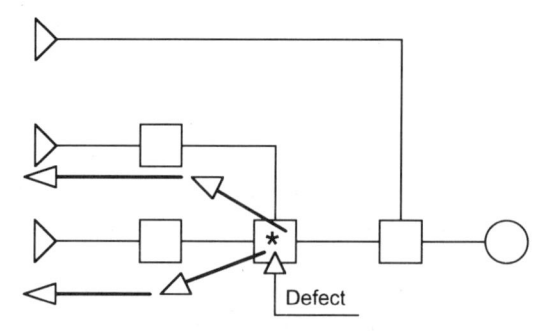

**Figure 9.8** Positive feedback on the production target of the bills of work, induced in the actions by the detection of a defective part

This kind of feedback is naturally impossible if the user doesn't want to *count* the amount of parts that have already been processed by each resource. The strategy may also induce a negative feedback on the production targets in certain cases which have been studied at the EPFL, depending upon the production objectives.

This powerful method takes many industrial situations into account with a full range of adapted reactions.

## 9.5 Conclusions

Sections 9.4.6 (i) and (ii) presented the new concepts introduced in the flow control module, and how it is integrated within the CAD method for industrial assembly.

The main flow control functions introduced in this section are listed in Figure 9.9.

| Function | Superv. | Resources | Switches | Purpose |
|---|---|---|---|---|
| Soft transit of BOW's | Execute | | | Process soft transit |
| | Interrupt | ⇐ Load | | Download BOW's |
| Asynchronous polling | Poll | | | Download prod. targets |
| | | ⇐ Reply | | Upload SQC local data |
| Manage production | | Execute | | Maintain local counters |
| | | | | Change production runs |
| | Reply ⇒ | Interrupt | | Deactivate |
| Manage destination | | Execute | | Manage routing rules |
| | Reply ⇒ | Interrupt | | Request routing rules |
| Manage traffic | Interrupt | | ⇐ Load | Download switching rules |
| | | | Execute | Choose & process rule |
| Manage user interrupts | Interrupt | ⇐ Reply | | Global request |
| | Monitor | Interrupt | | Local request |
| | | Execute | | Process interrupt |
| Manage resource interrupts | Reply ⇒ | Interrupt | | Request rule (routing) |
| | Reply ⇒ | | Interrupt | Request rule (traffic) |
| Execution | Global | Local | Local | |

**Figure 9.9** Summary of the flow control module functions

The industrial users appreciated the adaptation of the described functions to their expressed needs.

These tools, used both off-line within the simulated environment and on-line within the supervisor and controllers, provide a powerful answer to the challenge of today's needs for flexibility. By defining the same set of functions in these two worlds, the concurrent engineering approach is strongly promoted. Their proven feasibility and their standardisation allow the definition of marketable products.

# 10 Integration Aspects of the CAD Method

## 10.1 Introduction

CAD/CAM systems have been extensively used to streamline the process from piece part geometry design to CNC machining. The technology for generating NC machine cutting paths from surface models has matured over the last 40 years and finally received industrial acceptance. Assembly design and manufacture is still dependent on human experience and expertise. Assembly design begins at the very beginning of any product design process. It relates to the breakdown of the complete product into subassemblies. The development of the structure of the product defines the function and shape of the individual piece parts. Geometric details and features on each piece parts are refined until the designer is satisfied that the piece part could be assembled and deliver the functional requirements of the product. The assembly and manufacturing costs are directly linked to the design of the piece parts and the assembly.

Assembly design requires solid modelling technology. It is only in the late 1980s that solid modelling is sufficiently developed to be used successfully in large-scale engineering product design. Aerospace applications in the Boeing 777 and Rolls Royce Trent engine demonstrated the value of solid modelling to assembly design. Associativity between different geometric features in parametric and variational CAD systems provided the step change in CAD modelling concept from piece part design to that of the complete assembly of the product. This advance in CAD technology offers the opportunity for manufacturing companies to use CAD systems as their main repository of product and manufacturing data. The expertise and practice of performing product design, assembly planning and factory design could be systematically captured and integrated around the company CAD system.

## 10.2 Integrated Product Development Process

Competitive advantage in product design and development could be gained by the use of a CAD method to achieve an integrated product development. The economic benefits in using CAD/CAM systems to reduce draughting time is limited. The time and cost in updating a CAD file, redoing the dimensioning and plotting the drawing again is substantial. It is much more than just marking up a paper drawing. The ease of change and edit with a computer CAD model is restricted by the response time of the CAD operators and expensive plotters. The relationship between activities in a traditional design process can be seen in Figure 10.1.

Product design, assembly planning, factory design and tool design are all separate functions. There are no direct lines of communications between product design and the design of the shop floor and tooling. The capability and constraints of the factory floor are not effectively accommodated by the product design, leading to expensive production and assembly methods. Engineering changes may have to be done to the product design to ease difficulties in manufacturing.

To realise the full potential of concurrent engineering, all the

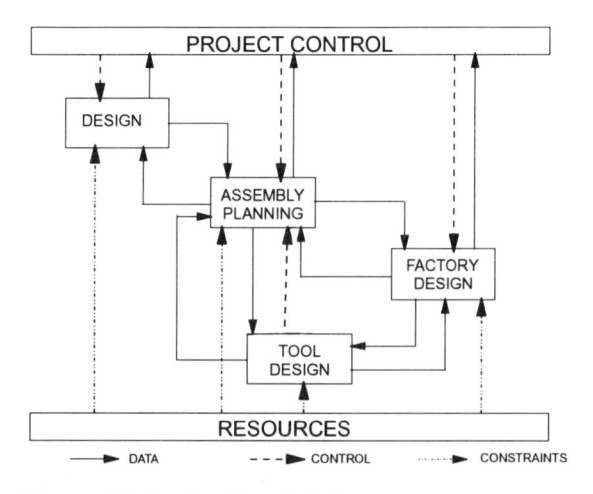

**Figure 10.1** Traditional design process

Integration Aspects of the CAD Method

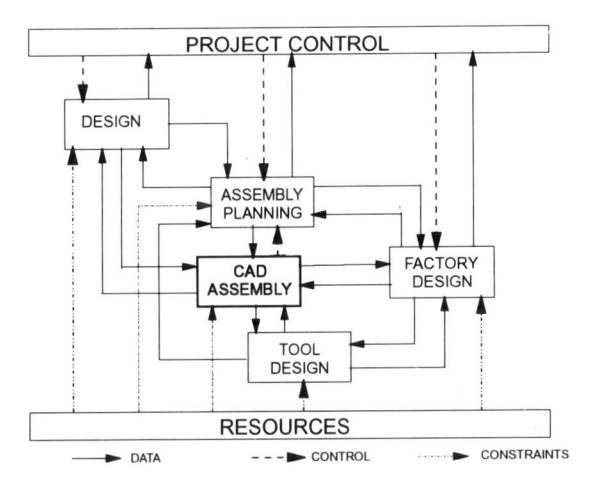

**Figure 10.2** Integration with assembly CAD

departments that have an impact on or are affected by the product design should have a part in the development of the product design. The participation of all relevant departments allows expensive design changes to be incorporated before drawing release. The assembly CAD system is the integration method of design and manufacturing functions. A process of full integration is in Figure 10.2.

The factory and tool design could be based on the same product information. This reduces any opportunities for errors caused by incorrect interpretation of drawings. Difficulties in factory and tool design could be reflected to the product and changes could be effected before the actual production. The integration permits the systematic concurrent design of product, processes and production equipment.

In situations where the manufacturing plant is capital intensive and the product design focus on derivatives from a basic design, the CAD system can be positioned closer to assembly planning (see Figure 10.3). The optimum positioning of the CAD assembly system in the design process is dependent on the product industry and company strategy.

Each activity in the process models is performed by engineers who are expert in their work. Their knowledge and expertise could be captured and enhanced by decision support systems. In the CAD method for industrial assembly, systems are specified that support each activity. The product design function is

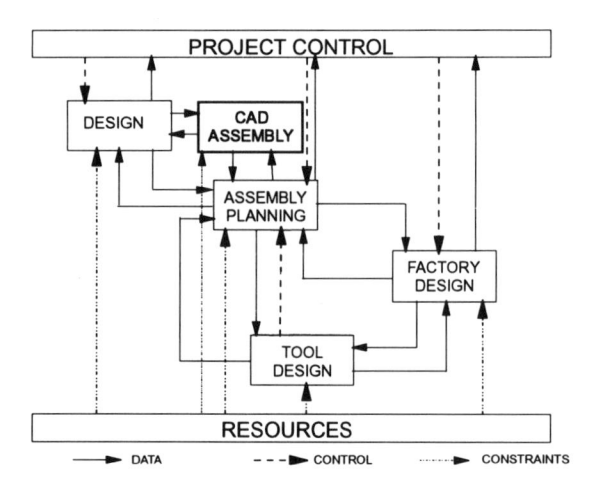

**Figure 10.3** CAD positioned closer to assembly planning

supported by the product design for assembly module. The assembly planning function is supported by the assembly planning module. The factory design and tool design is supported by the resource planning module.

## 10.3 Product Data Integration

The integrated CAD system provides the platform for simultaneously performing design of the product, assembly processes and equipment. To reduce the product development time, overlaps of design activities requires process and equipment design to start before the product design is completed. Through interaction of product and process design, a more competitive product could be developed in a shorter time to market. The detail product and assembly data and their relationships defined in the CAD method for industrial assembly is used as illustration.

### 10.3.1 Product Design For Assembly

The aim of the product design for assembly module is to provide positive advice to the user. The DFA module consists of the five submodules: product structure; handling; feeding; positioning/insertion; and joining. These might be accessed during any stage of the design process. The analysis is designed to support the different stages of design, from concept through to detail.

At each stage and as the analysis progresses, different combinations of data, with increasing levels of detail, are required. If insufficient data is available the user will be requested to provide it. The more data that is available from the CAD system and other modules the less user involvement is required.

The complete set of input data required by the DFA module is listed below in the order in which the data is likely to be obtained.

1. specification;
2. bill of materials;
3. component/assembly functionality;
4. standard components/assemblies;
5. links with variants;
6. precedence constraints;
7. assembly tree;
8. selection of base component;
9. component/assembly liaisons;
10. motion between components/assemblies;
11. geometry;
12. tolerances;
13. surface functionality;
14. component material;
15. joining processes;
16. resources available.

Each submodule requires different combinations of this data. The required data and the involved outside modules are shown in the following table. In all cases the user will be requested to provide an input if insufficient data is available from another source. The modules may operate with all or none of this data, or any combination of the data. Obviously the more data that is available the less user input is required.

| Module | Data required | Other modules involved |
|---|---|---|
| Product Structure | [1],[2],[3],[4],[5],[7] [8],[16] | Assembly Planner, Resource Planner |
| Handling | [11],[12],[13],[14],[16] | CAD, Resource Planner |
| Feeding | [11],[12],[13],[14],[16] | CAD, Resource Planner |
| Positioning & Insertion | [9],[10],[11],[12],[16] | CAD, Assembly Planner, Resource Planner |
| Joining | [9],[10],[12],[13],[14] [15],[16] | CAD, Assembly Planner, Resource Planner |

In addition to the above, methods time measurement (MTM) data is used to provide timing and costing data for manual operations, whilst the costing data from the resource planner is used for the automated operations.

## 10.3.2 Assembly Planner

The goal of the assembly planner is to produce an assembly plan to be sent to the design or resource planner modules.

Recall that the set of submodules involved in the Assembly Planner is:

(I) enhancement of the product model;
(II) computation of precedence relations;
(III) computation, evaluation and selection of a set of assembly plans;
(IV) creation, edition and storage of an assembly plan.

Recall also that the set of data involved in the Assembly Planner is:

1. product model (CAD): BOM, etc.;
2. enhancement of product model (needed for automatic scenario);
3. precedence relations;
4. set of assembly plans;
5. the final assembly plan.

The minimum set of data is [1], in other words, the product model (BOM, etc.). The minimum set of submodules to use is [IV], in other words, the assembly plan editor. The combination { [1], [IV] }, called "step number $a$" in the following array, represents the basic tool needed to build an assembly plan. It has been called "manual scenario" in Chapter 5. Then we finally obtain [5], the desired assembly plan.

The main available combinations of data and submodules are shown in the following table:

| Step | Involving data: | Using submodules: | We obtain |
|---|---|---|---|
| a | [1] | [IV] | [5] |
| b | [1], [3] | [III], [IV] | [4], [5] |
| c | [1], [2] | [II], [III], [IV] | [3], [4], [5] |

Explanations:

- The assembly plan is built by hand (manual scenario).
- The operator inputs by hand all precedence constraints (geometry, technology, user knowledge)—item 3. Then the module can automatically compute, sort and select a set of assembly plans—items 4, 5.
- Knowing the enhanced product model—item 2, the module can automatically compute the precedence constraints—item 3—and after that, a set of assembly plans.

In other words, missing data do not affect the assembly planner efficiency. The only condition is that the user *manually inputs a few precedence constraints* for substituting data the system cannot find in the CAD database. But, if precedence constraints are missing, results generated by the assembly planner can be unpredictable. Many trees will give spurious assembly sequences. Therefore, we can finally say that if data are missing, the user has to care about the substituting precedence constraints.

## 10.3.3 Resource Planner

The resource planner receives as principal input (from the assembly planner), the list of operations and the precedence constraints between these operations. The resource planner is also linked to a database of equipment that is related to the operations.

### (i) Manual planning

As soon as some operations are designed, even if those operations are not ordered, the resource planner, through its related database of equipment, allows the user to verify if equipment for this operation is available, and if the cost is reasonable. The choice of costly equipment (if only costly equipment is suggested) or the inability to reuse or recycle some equipment gives indications of problems in the design of the product or the assembly plan. This shows clearly that very early feedback is possible with other modules, even if the model is still incomplete.

The first goal of the resource planner is to produce a logical layout, that is, regrouping the operations on cells or stations, attributing sets of equipment to each of them, while being constrained by a user-given cycle time.

As soon as the operations are defined or suggested by the user with the help of the DFA module and held in the assembly planner (in varying possible levels of detail, as stated in the previous section), the user may begin work with the resource planner. A specific workshop configuration may be chosen against a global cycle time (related to required volume of production and marketing-related data) and regrouping operations on stations may be carried out. Operation durations may be estimated by the user or inferred from equipment already chosen, and the method will then provide a plan of the operations according to the workshop information just chosen and the cycle time constraint. The resource planner might also be able to provide the minimum cost of equipment for each operation, provided that some equipment restriction has begun.

At this point, as no precise costs or duration are given, and as an incomplete set of operations is offered, no optimisation of the layout may be conducted. However, the manual resource planning tool may already be used with relevant first results. The user may already have an indication of the accuracy of his or her workshop configuration choice or the possible number of similar workstations to ensure the required volume of production.

In the design of the physical layout, the user may already choose some equipment with his or her limited knowledge of the required operations, and have first estimates of space constraints, which, in turn, might allow a more accurate choice of equipment.

**(ii) Equipment selection**

Predefined attributes are matched with the equipment in the equipment database to select the appropriate ones.

The choice of possible equipment applicable to each operation is reduced by *gradual restriction*. The more attributes are given, the more precisely the equipment is specified, and the less equipment from the equipment database is proposed. Thus not all the attributes are required to be specified. This will simply have the effect of enlarging the choice of possible equipment.

There is a minimum information requirement: the kind of operation (e.g. welding, screwing, . . .) must be provided. If an operation is completely *un*specified, *any* equipment could in theory carry out the operation.

The attributes have been divided into those which are *process-related* (i.e. they specify the operation) and those that are *equipment-related* (i.e. setting specific requirements on the equipment).

Once the kind of operation (i.e. the most basic process-related attribute) is specified, the relevant equipment-related attributes are known, and can serve to narrow the choice of equipment even when not all the process-related attributes are fixed. The DFA module can advise at the process selection stage.

The optimisation step takes the lists of possible equipment for each operation and selects for each operation *one* equipment in such a way that the total cost of the line is near-optimal.

To be able to evaluate the merit of each assignment of equipment to operation, the algorithm has to be able to compute the duration of the operation and its cost, given an equipment. However, in the case where the operation is not sufficiently defined, i.e. if the *only* data for an operation is that it is screwing, it is difficult to compute how long it will take, as that figure depends on the kind and the dimensions of the screw.

The output of the resource planner, i.e. a logical layout and (through a physical layout editor) a physical layout is supplied to the simulation module. As soon as the set of stations are designed with some chosen equipment, an incomplete but useful process model may be given to the simulation module.

**(iii) Data requirement**

1. With a set of operations
   (a) *Tool capabilities*. As the joining analysis within the design for assembly module is run, the process information contained within the DFA module is used in conjunction with the equipment database to guide the designer in selecting the most suitable process for which equipment is available within the company.
2. With an assembly plan (operations + precedence constraints) and a cycle time:
   (a) *Choice of equipment*. If no, or only very costly, equipment is available feedback may be given on the difficulty of realising an operation. The redesign and assembly planning modules can then be re-run to solve these difficulties.
   (b) *Grouping of operations*. Using basic logical layout design capacities, a check is made on the cycle time and workshop configuration constraints. Multiple instances of workstations can be created by the resource planner in order to satisfy the constraints. First data are provided to the physical layout editor.

(c) Line balancing with chosen equipment, or by default, time estimates.
3. With an assembly plan, accurate database of equipment, and a cycle time. The user has full automatic planning capacities. He/she has detailed information on the costs and durations of the operations. The line can be optimised with the help of the resource planner's algorithms (see Section 6.4.3(iii)).

### 10.3.4 Simulation

The simulation module is used for the testing of a production system that has been designed on the basis of static planning data under the conditions of a real-time run.

The production system is represented in the simulator by dynamic data and process models. The complete model is built up from single models of the elements of the production system, which include substations, workers and material flow devices.

Products and operations are assigned to the elements of the production system. Orders, scheduling and flow control represent the dynamic factors in the simulation model.

The simulator is an evaluation tool in the concurrent engineering approach. It is used to find out whether a planned production system works as planned in the dynamic run. Bottlenecks or unsatisfactory capacity load that are discovered in the simulation can be overcome by adjusting the production system. Thus simulation is a very useful tool in the off-line planning. Furthermore, the generated dynamic data are used to validate the solutions provided by the off-line modules before implementing the real assembly system. The decisions of the off-line modules are influenced by the simulator. It supplies also the flow-control module with a simulated system in order to enable the user to complete flow rules before they are transmitted to the real layout.

Generally there are three ways of using the simulator, not only at the end of the planning process, but in parallel. These are explained in more detail below.

**(i) The use of hierarchical models**

The model of the production system can be built in a hierarchical structure. This means for example a cell of the system can be managed as a cell in the simulator, if the single processes do not need to be further detailed for evaluation purposes. If required the cell can be refined into stations.

# Integration Aspects of the CAD Method

|  |  |
|---|---|
| **(ii) The use of partial models** | The simulator can also work with incomplete models of the production system, providing partial information only. For example, if the qualification of workers is of no interest for the operations to be performed, this does not need to be specified in the simulator. |
| **(iii) The use of "raw data"** | The "raw data" simulation model is based on a very simple set of elements with a very small set of important key data. Therefore, the use of such a simulator is easy and fast. It can be used in a very early state of the planning of the production system, e.g. for a first check of the capacity values. |
| *10.3.5 Scheduling* | The scheduling module has five separate phases (Figure 10.4). The first two are required for the definition of user defined |

The data structures for the cell have to be the same set as for a single station. Of course, the data cannot be as accurate as if single stations are considered.

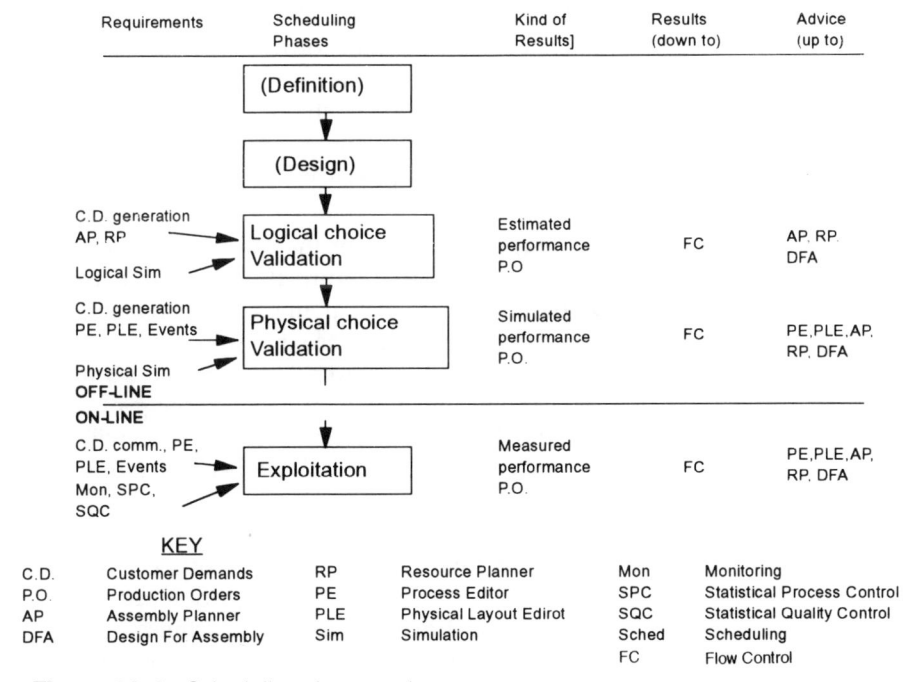

| Requirements | Scheduling Phases | Kind of Results] | Results (down to) | Advice (up to) |
|---|---|---|---|---|
|  | (Definition) |  |  |  |
|  | (Design) |  |  |  |
| C.D. generation AP, RP Logical Sim | Logical choice Validation | Estimated performance P.O | FC | AP, RP. DFA |
| C.D. generation PE, PLE, Events Physical Sim OFF-LINE | Physical choice Validation | Simulated performance P.O. | FC | PE,PLE,AP, RP, DFA |
| ON-LINE C.D. comm., PE, PLE, Events Mon, SPC, SQC | Exploitation | Measured performance P.O. | FC | PE,PLE,AP, RP, DFA |

**KEY**

| | | | | | | | |
|---|---|---|---|---|---|---|---|
| C.D. | Customer Demands | RP | Resource Planner | Mon | Monitoring |
| P.O. | Production Orders | PE | Process Editor | SPC | Statistical Process Control |
| AP | Assembly Planner | PLE | Physical Layout Edirot | SQC | Statistical Quality Control |
| DFA | Design For Assembly | Sim | Simulation | Sched | Scheduling |
| | | | | FC | Flow Control |

**Figure 10.4** Scheduling data requirements

methods and criteria and are necessary to allow the programming of the user-defined heuristics and the choice of a heuristic, according to a defined set of criteria. Scheduling can start at the beginning of the design process, with the first two phases. The module will provide, via' the man–machine interface, all the necessary information.

1. The first phase is the definition of scheduling criteria to be taken into account during scheduling, and the formats for the programming of user defined heuristics, if necessary.
2. The second phase is the actual creation of any user-defined heuristics, if necessary.
3. Scheduling can only begin once the module has at least a logical layout. The result of this scheduling, together with the logical simulation, will give an estimated performance, which can be validated by means of comparison with other methods, by means of rough simulation.
4. Once the physical layout is available, the module calculates the simulated performance, and allows its validation by comparison with other methods, by means of detailed simulation.
5. The chosen heuristic is now ready for on-line application. The module obtains a list of customer demands or production orders and transforms them into bills of work for the resources. For each new order, the module obtains the necessary information from the process and physical layout database, which result from the interaction of the user with the process and physical layout editors.

### 10.3.6 Flow Control

The design cycle time of the flow control activity is to be separated in five phases (Figure 10.5). The first two are required for the definition of user defined methods and criteria and are necessary to allow the programming of user-defined rules and routing strategies, according to a defined set of criteria.

Flow control can start at the beginning of the design process with the first two phases. This module will provide, via the man–machine interface, all the necessary information for these first two phases.

1. The first phase is the definition of the flow control criteria to be taken into account during flow control (availability of operators, breakdowns of machines, etc.), without any knowledge of a particular layout. It also provides the formats for programming of user-defined rules and strategies.

# Integration Aspects of the CAD Method

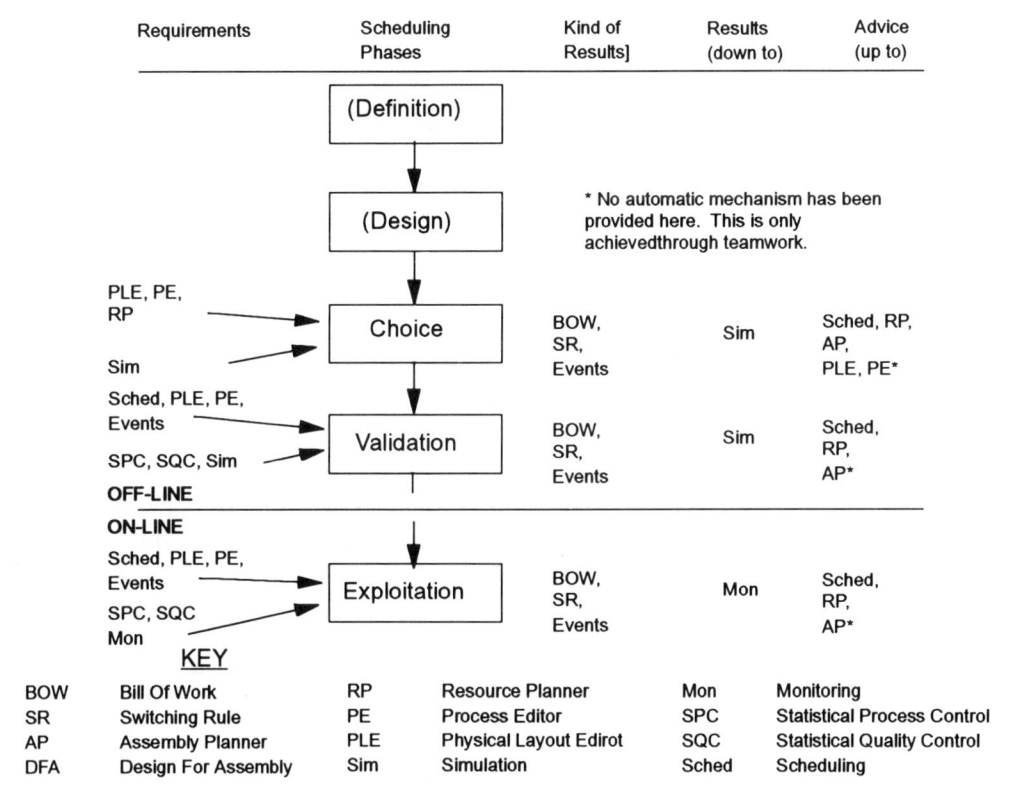

|   | Requirements | Scheduling Phases | Kind of Results | Results (down to) | Advice (up to) |
|---|---|---|---|---|---|
| | | (Definition) | | | |
| | | (Design) | | * No automatic mechanism has been provided here. This is only achieved through teamwork. | |
| | PLE, PE, RP<br>Sim | Choice | BOW, SR, Events | Sim | Sched, RP, AP, PLE, PE* |
| | Sched, PLE, PE, Events<br>SPC, SQC, Sim | Validation | BOW, SR, Events | Sim | Sched, RP, AP* |
| **OFF-LINE** | | | | | |
| **ON-LINE** | Sched, PLE, PE, Events<br>SPC, SQC<br>Mon | Exploitation | BOW, SR, Events | Mon | Sched, RP, AP* |

**KEY**

| | | | | | |
|---|---|---|---|---|---|
| BOW | Bill Of Work | RP | Resource Planner | Mon | Monitoring |
| SR | Switching Rule | PE | Process Editor | SPC | Statistical Process Control |
| AP | Assembly Planner | PLE | Physical Layout Edirot | SQC | Statistical Quality Control |
| DFA | Design For Assembly | Sim | Simulation | Sched | Scheduling |

**Figure 10.5** Flow Control data requirements

2. The second phase is the actual creation of any user-defined switching rule or routing strategy, as well as compensation strategies when encountering quality problems leading to the rejection of bad parts.
3. Flow control begins once the module has at least a physical layout, since a logical layout does not define a way of controlling flow, due to the lack of buffers and switching nodes. Flow control may begin sooner if the concepts of the traffic network, within the logical layout are used. This enables consideration of the precedence constraints between the assigned resources. The result of the flow control, together with either the logical layout or the physical layout, is a set of flow control data (rules, strategies, user-defined heuristics) that can be validated by means of simulation.
4. Once the physical layout is available, the module interacts

with the simulation module by means of events, thus leading to its validation.
5. The chosen flow control data are now ready for on-line application. The module obtains events, system state and system history (SPC, SQC) from the monitoring module and generates changes within flow control data as well as events to be processed by the monitoring and resource controllers.

## 10.4 Man–Machine Interface

To support the sharing of data in an integrated CAD method for assembly design, the man–machine interface (MMI) needs to present the product geometry, assembly structure and the factory layout information to support the different modules (see Figure 10.6).

The geometry window supports the building up of the product using solid model. The geometry of the assembly and piece parts can be sketched, edited, deleted and manipulated. The assembly planning window displays the assembly plan. The assembly actions could be selected and linked together to form the assembly plan (see Figure 10.7).

The assembly plan editor supports the creation of the assembly plan automatically or manually. The plan could be

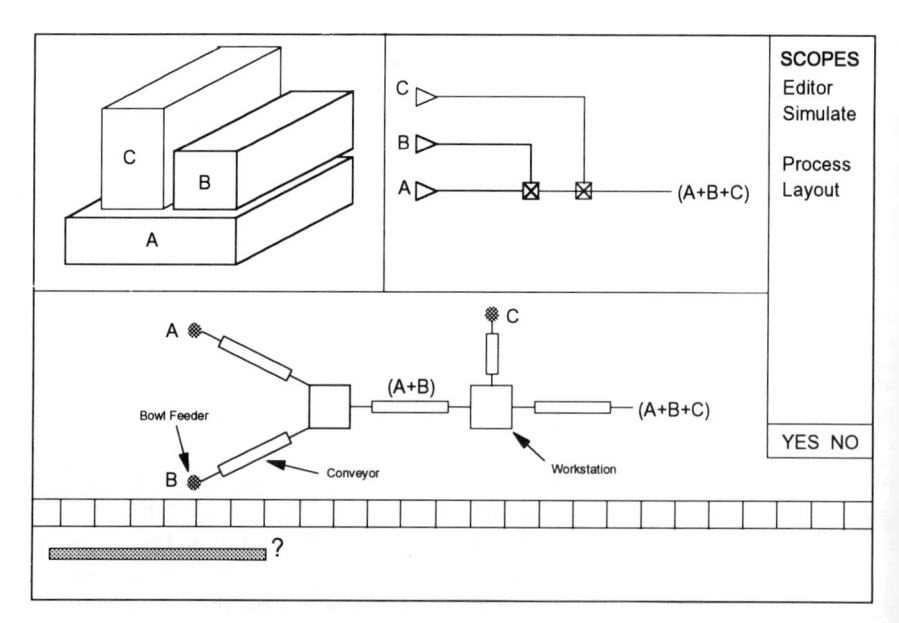

**Figure 10.6** Integrated man–machine interface

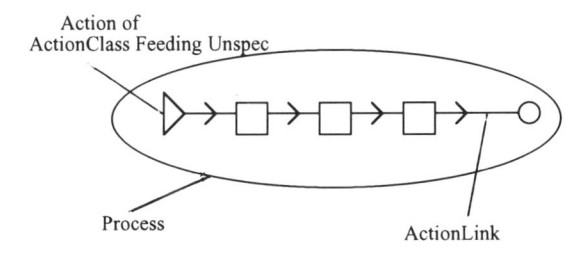

**Figure 10.7**  Assembly process plan

modified and redrawn during analysis of impact to product design and factory design. The piece parts in the geometry window are linked with the feeding actions in the assembly plan. The assembly actions are linked with the joining relationships of the piece parts. Pointing at the feeding action in the assembly plan window causes the corresponding piece part in the geometry window be highlighted. Pointing at a joining action in the assembly plan window causes the sub-assembly in the geometry window to be highlighted. This association between product and assembly plan is a crucial part in supporting the concurrent design of the product and the assembly plan.

The factory layout window displays the layout of the physical resources to assemble the product. The physical workstations are linked to actions in the assembly plan (see Figure 10.8).

The physical layout editor supports the user in the selection of physical resources and the modification of the workstation layout. Additional buffers and conveyors can be added to the physical layout. The links from the physical resources extend from the assembly plan to the product model. Production flow simulation can be performed to evaluate the goodness of the layout.

This integrated man–machine interface enables the concurrent design of the product, assembly plan and factory layout. The flexibility in editing and changing any one of these encourages the design team to test more product design concepts and evaluate their impact on the assembly line. From the product design point of view, the initial assembly plan could be linked to a sketch concept of the piece part. Upon satisfactory selection of assembly process and factory design, details of the piece part can be developed to make best use of the assembly information. Features for joining, handling and feeding could be added to

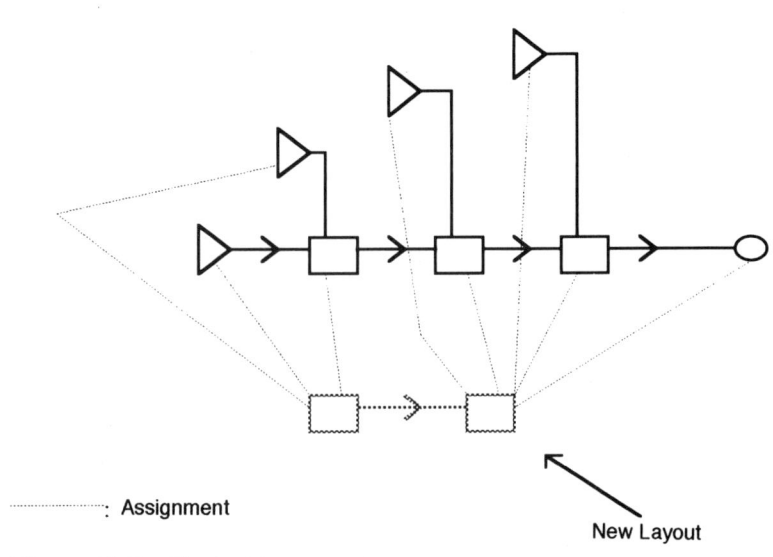

**Figure 10.8** Physical layout and assembly plan

optimise the design for manufacture. The design for assembly analysis can be invoked according to the context of application. Different analyses are enabled depending on the way the user selects components. Part-related analyses are enabled when one or more components are selected. Assembly level analysis is only enabled when two or more components are selected.

| | |
|---|---|
| 0 component or action selected | Product structure analysis |
| 1 component selected | Product structure analysis<br>DFA analysis<br>Create actions set<br>Insert actions set<br>Modify actions set |
| 2 or more components selected<br><br>1 action selected | Product structure analysis<br>Create actions set (in order of selection)<br>DFA analysis<br>Modify action |

The integration of geometry, process plan and factory layout on the same display screen supports the concurrent considera-

Integration Aspects of the CAD Method   253

tion of factors affecting the three sets of information. In industrial design, teams of engineers participate in the design. Network computing and computer supported design conference are additional technology to support concurrent team design.

## 10.5 The Man–Machine Interface in CATIA

The following figures show the MMI as it was implemented in the CATIA CAD system.

Figure 10.9 shows an exploded view of one of our industrial case studies: an electrical contactor. This is the standard geometry screen for solid modelling.

Figure 10.10 proposes a view of an assembly plan associated with the above-mentioned product. The assembly process editor is displayed in the bottom part of the screen. The assembly plan of the contactor is being developed. The main sequence of the assembly plan is represented by the linked line of boxes. The symbols in the box represent different types of process actions.

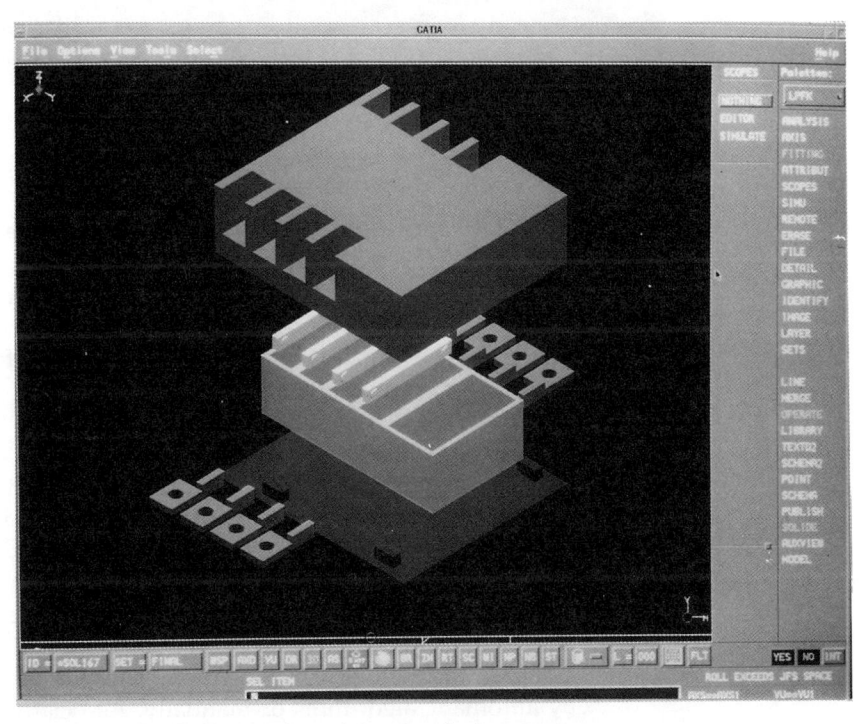

**Figure 10.9**  Exploded view of the electrical contactor

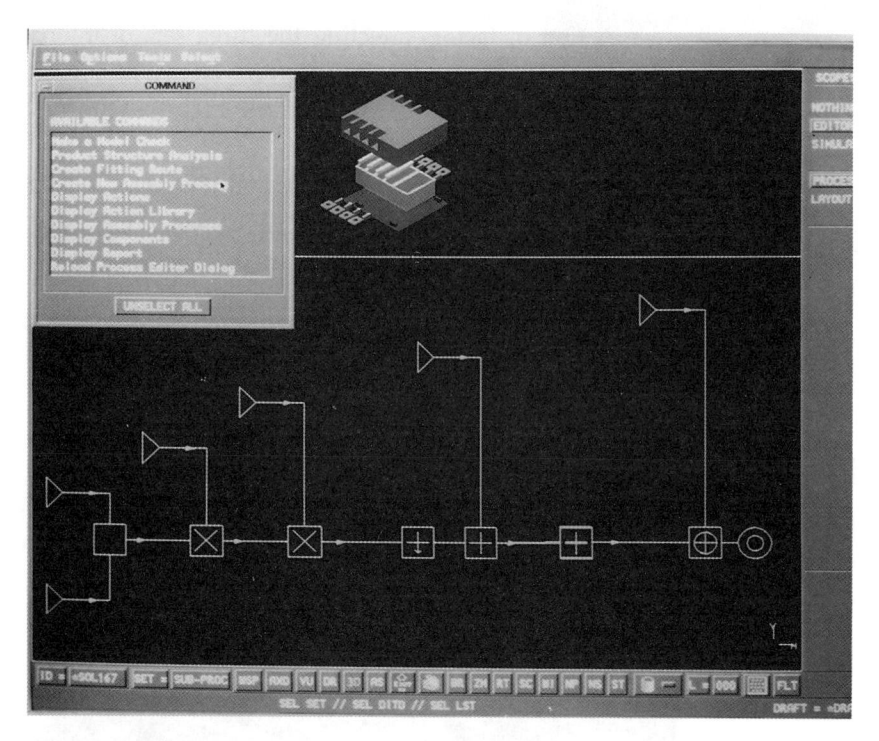

**Figure 10.10** Assembly plan

The last operation, represented by a double circle, is the delivery of the whole assembly. The triangular symbols represent feeding actions. Each feeding action is linked to a part in the solid model. Associativity between the parts and the process model is fully maintained. The user can verify the stage of assembly by selecting a process box. The parts that have been assembled at that stage will be highlighted in the solid model screen. The pop-up menu box displays the context-sensitive range of commands that the user can select. Figure 10.11 displays the editing commands concerning the different actions of the plan. This panel could be customised for the user company.

Figure 10.12 shows the logical layout of the assembly line after the resource planning action. The assembly processes are assigned to the logical workstation. The line could be balanced by automatic algorithms or manually. Associativity between the workstations, processes and parts is fully maintained. The user

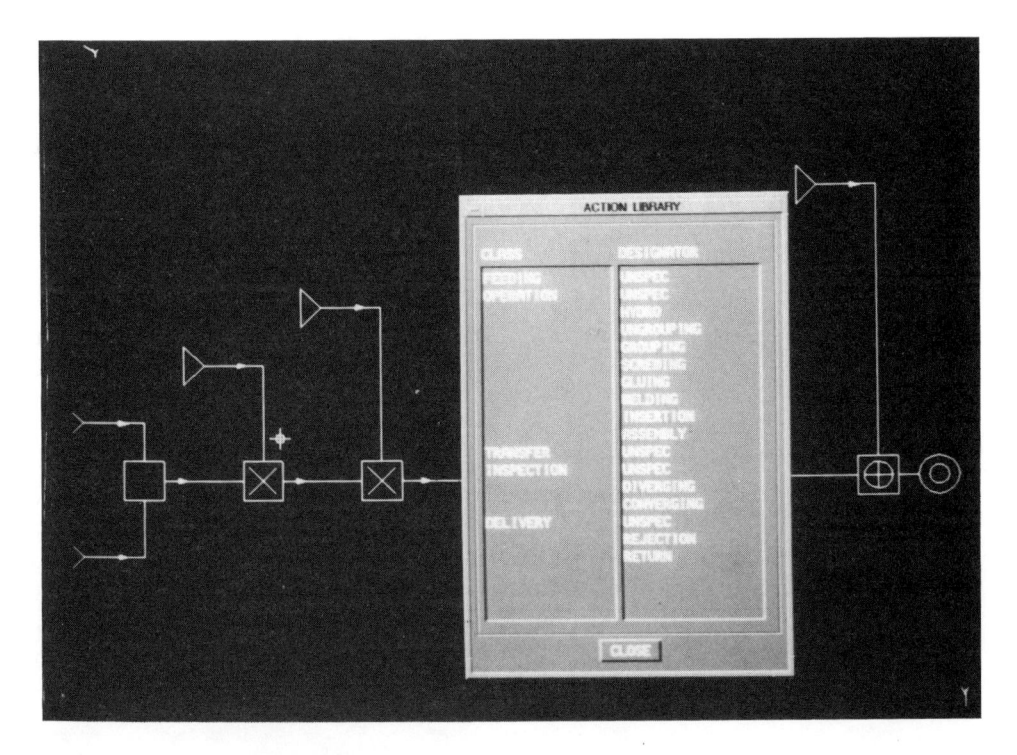

**Figure 10.11** Editing of actions

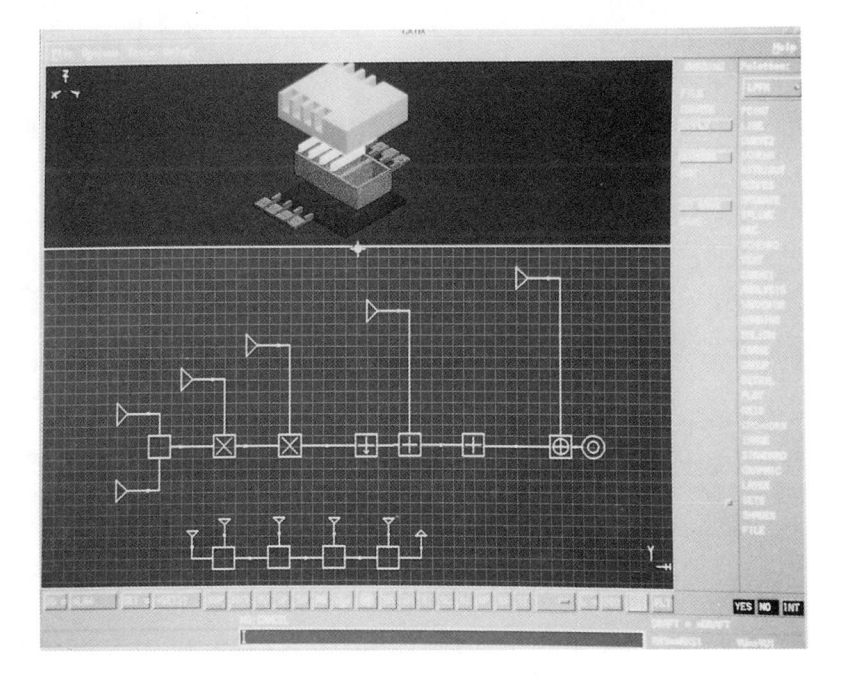

**Figure 10.12** Logical layout of the line

can check the status of the part geometry, assembly process and resource layout. The common display and data architecture supports the user to refine the product design with the assembly and shopfloor situation fully represented. DFA analysis could be called from the workstations, process boxes or geometric parts. The assembly and the parts will be evaluated for ease of assembly and suitability of equipment. Recommendations for redesign could be provided to assist the design user.

Figure 10.13 shows the result of the physical layout editor, conveyors, buffer stations and switching stations are added to fully represent the physical layout of the assembly line. As shown on the layout, the user decided to build up its line out of conveyors. Two derivations allow the flexible control and optimal utilisation of resources. In order to start the simulation module, the user now has to define production orders or can use the list of production orders generated from the scheduling module. In comparison with a traditional simulator, there is no need to build up further simulation modules; this integrated

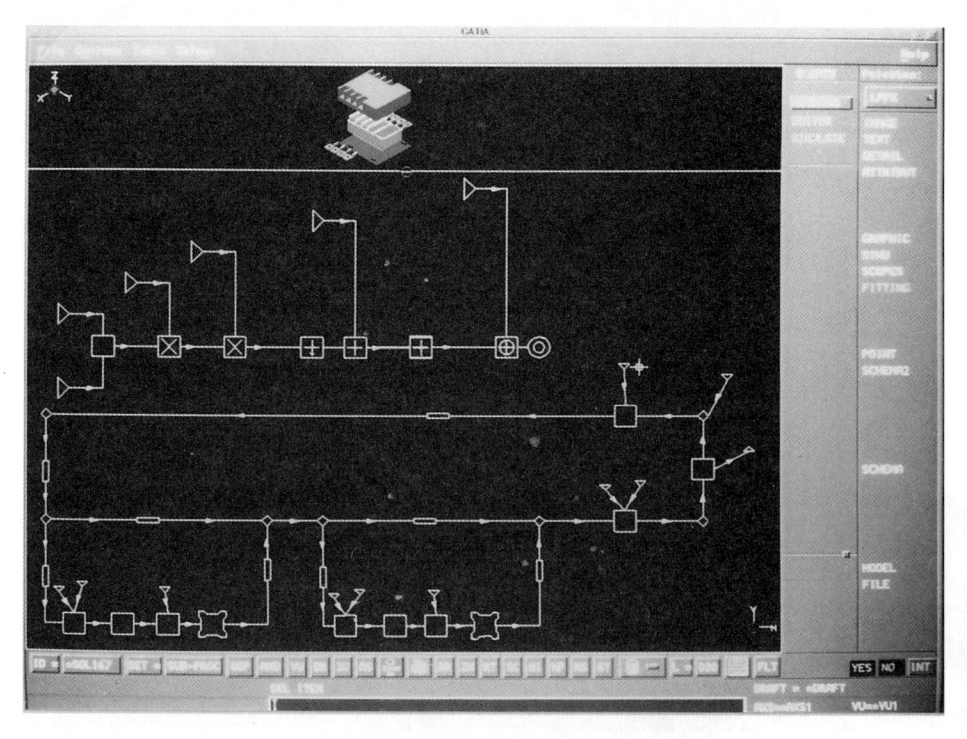

**Figure 10.13** Physical layout of the line

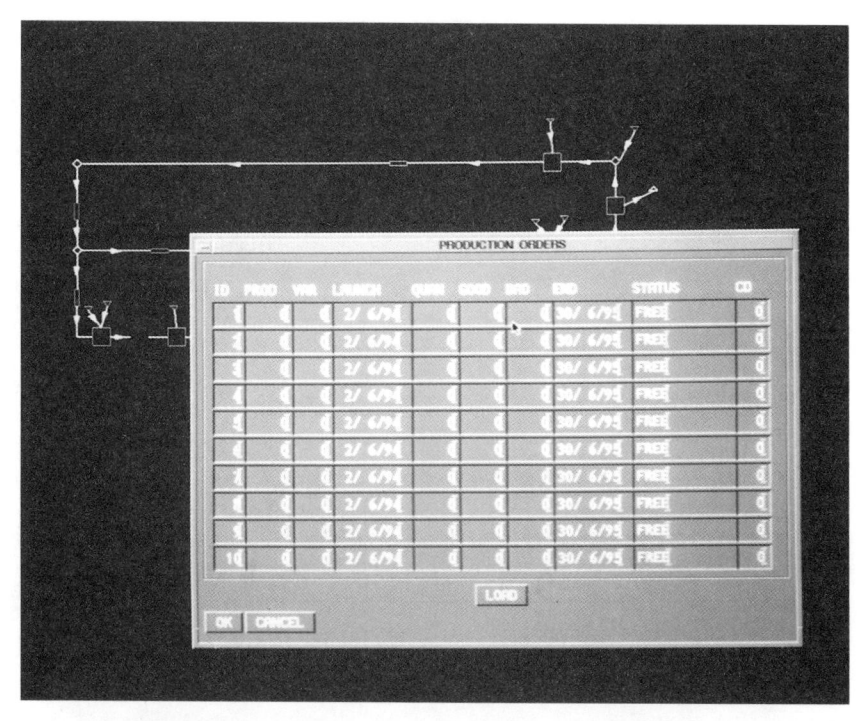

**Figure 10.14** Simulation and production order panel

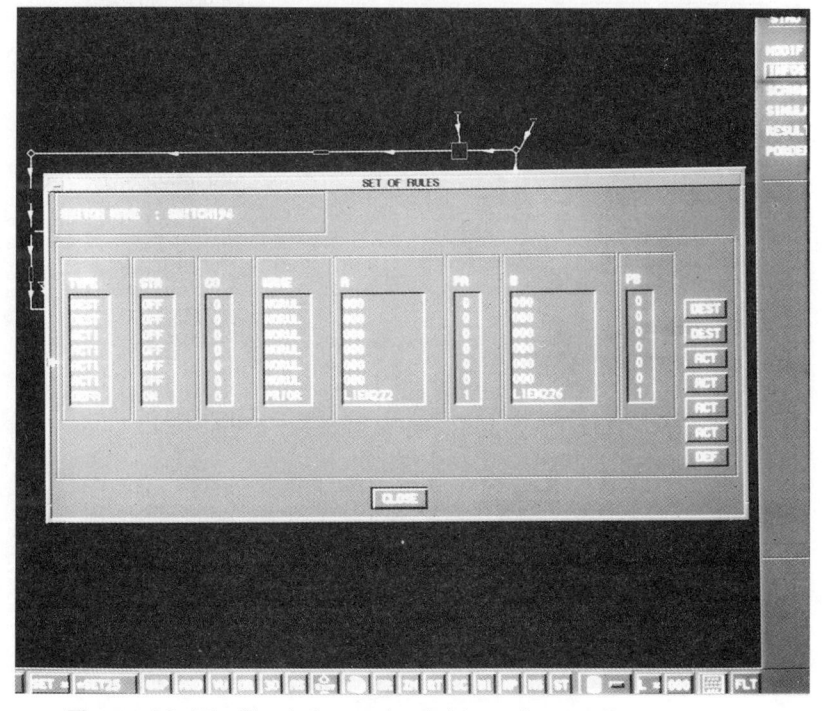

**Figure 10.15** Simulation and switching rules panel

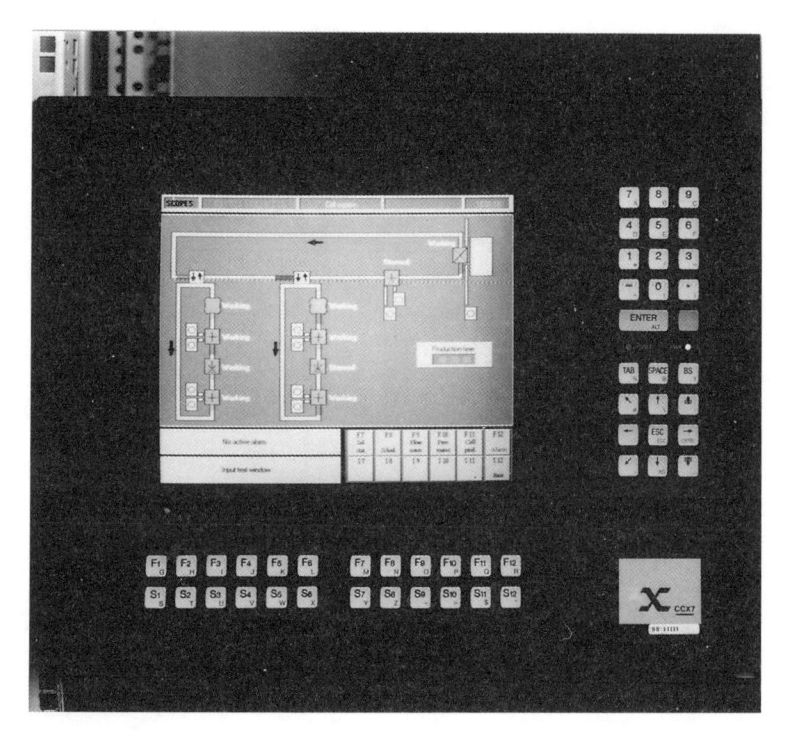

**Figure 10.16** Cell supervisor

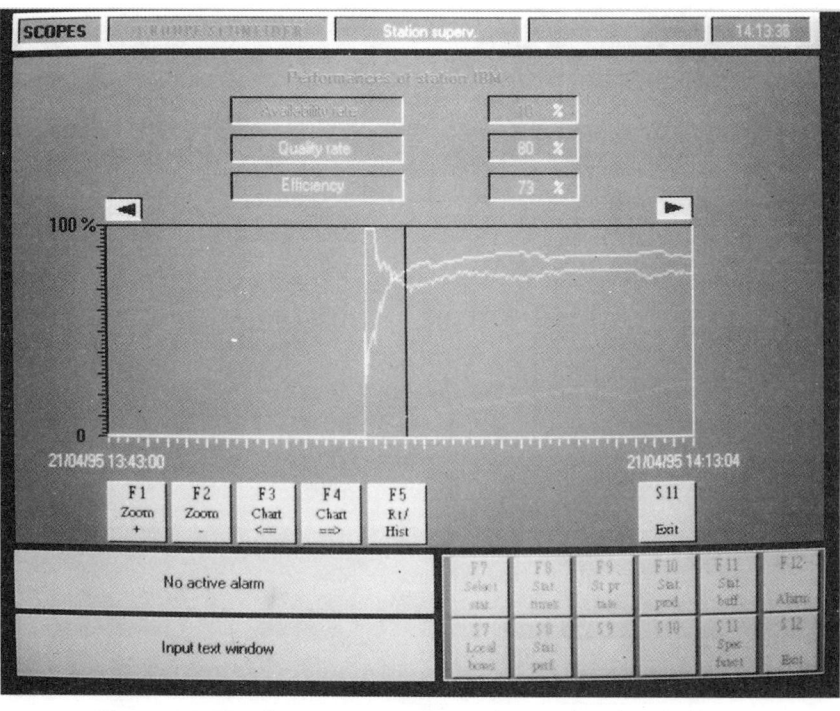

**Figure 10.17** Station supervisor: performances of a station

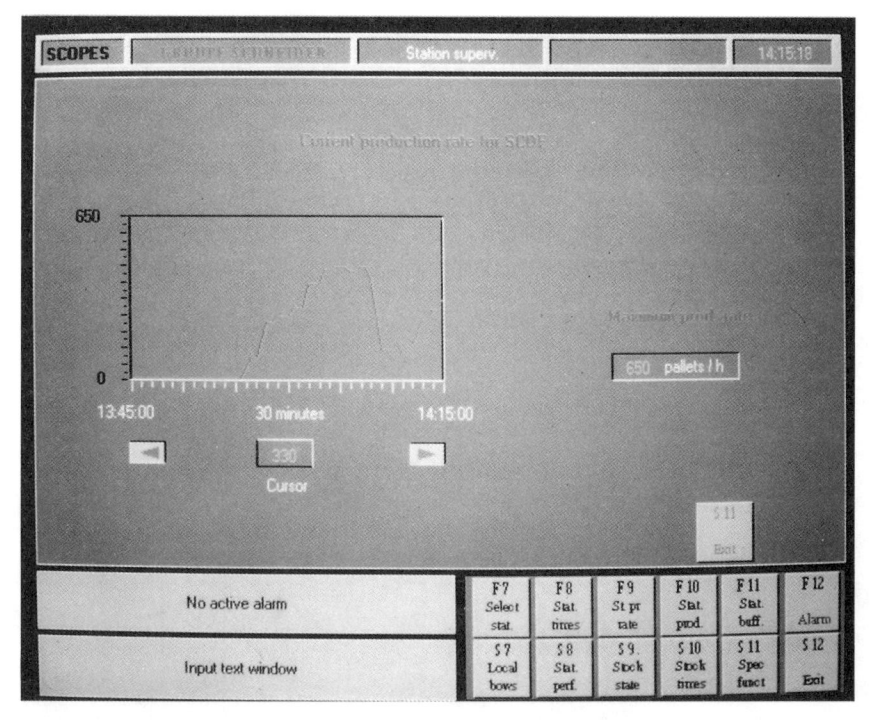

**Figure 10.18** Station supervisor: production rate

**Figure 10.19** Scheduling module: customer demands

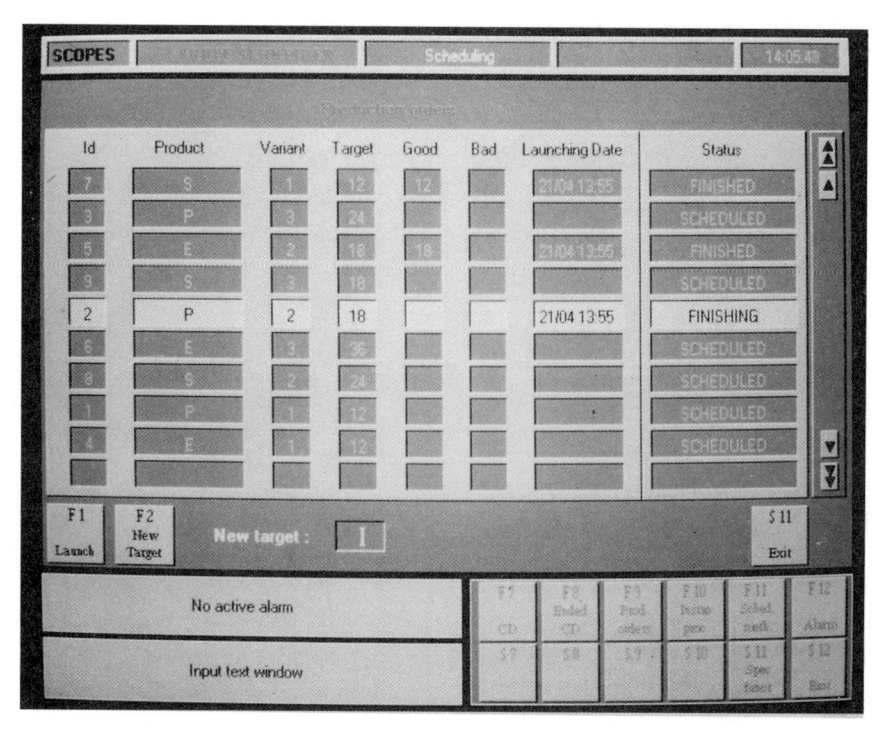

**Figure 10.20** Scheduling module: production orders

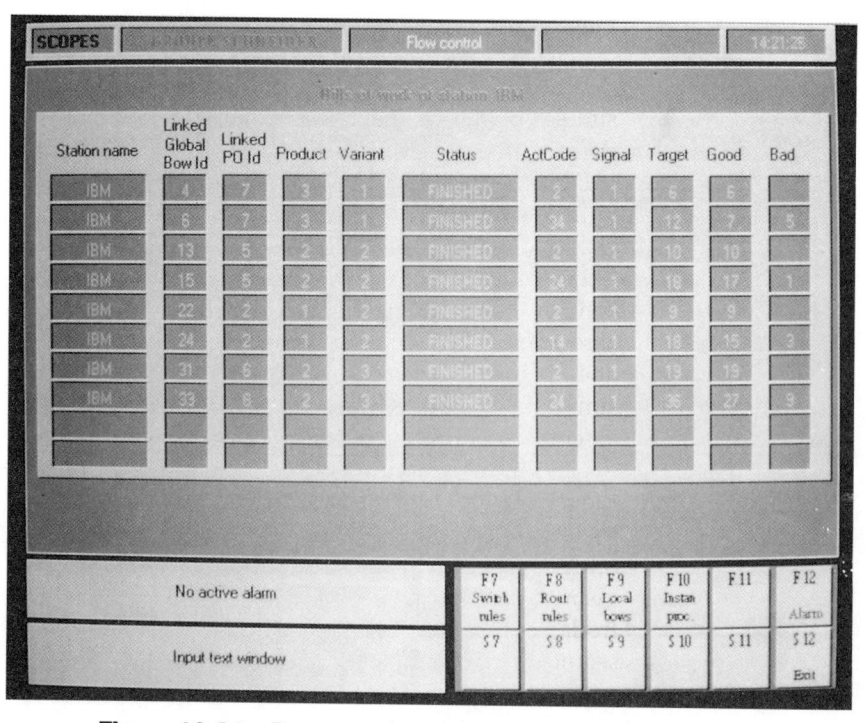

**Figure 10.21** Flow control module: bill of work of a station

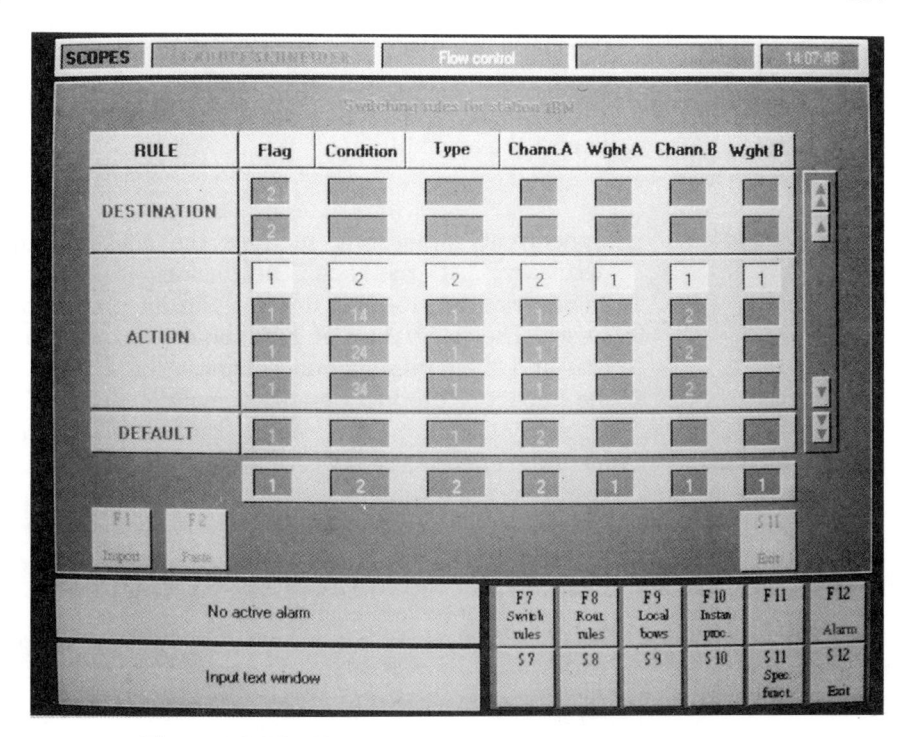

**Figure 10.22**  Flow control module: switching rules

simulator allows the user to start a simulation straight after defining the physical layout.

Figures 10.14 and 10.15 show the simulation module running with the production orders and the switching rules panels. The expected production orders can be easily entered into the simulation module. The user can try different switching and flow control rules to evaluate the best strategy to control the assembly line. Different equipment and resource breakdown and recovery scenarios can be played out. The results of the simulation could be used to improve the physical layout, the assembly plan or even the product design.

The following figures presents the line controller: Figure 10.16 the cell supervisor, Figures 10.17 and 10.18 some functions of the station supervisor, Figures 10.19–10.22 some displays of the on-line scheduling and flow-control modules. The cell supervisor is an enhanced Modicon 77 programmable logic controllable that has been specifically adapted for our work. A seamless information transfer protocol had been established to transfer the scheduling and flow-control rules from the CAD station to the cell supervisor. Monitoring functions in the supervisor ensure rapid fault diagnosis and high production utilisation of the assembly line.

## 10.6 Conclusions

Concurrent engineering requires the effective support of the necessary CAD methods. The business benefits in concurrent engineering are realised through the integration of design processes. The integration of product and manufacturing information and their presentation enhances the explicit evaluation of the impact of product design on manufacturing. This integrated assembly CAD system forms the hub of product data management in concurrent engineering organisations.

## 10.7 Bibliography

Sharpe, J. and Oh, V. (1994) Computer aided conceptual design, *1994 Lancaster International Workshop on Engineering Design*, Lancaster University Engineering Design Centre.

Starkey, C. V. (1992) *Engineering Design Decisions*, Edward Arnold, London.

Ullman, D. G. (1994) *The Mechanical Design Process*, McGraw-Hill, New York.

# 11 *Introducing the Integrated CAD Method into Companies*

## *11.1 Introduction*

The integrated CAD method facilitates the practice of concurrent engineering by providing the single platform for data and process definition. Through this single definition, communication between design and process engineers could be conducted in an unambiguous manner. The activities of product design, process planning and factory design could be supported by design for assembly, assembly planning, resources planning and simulation analysis. The full business benefits of the integrated method depends on a successful implementation programme.

Early generation CAD systems are difficult to use and require a high level of computer literacy. This has led to specialist CAD draughtsmen being used to create the CAD drawings from instructions by design engineers. Price reductions in CAD hardware and software results in rapid penetration of CAD systems in industry in the 1990s. However, a recent survey (Bostock Marketing Group, 1990) reported only 5% of organisations surveyed believed they had achieved the anticipated benefits.

The proposed integrated CAD method for assembly enables the sharing of product and process data. Concurrent design of product and process requires a new way of work and organisation. Communication between the different members in the design team to achieve a competitive product to market is the key element of success. Team members have to satisfy multiple objectives and meet different constraints subject to the uncertainty about the market. Despite the best of intentions, conflicts are inevitable. Effective concurrent engineering systems help to avoid conflicts due to the lack of understanding.

The human element is the most critical factor in introducing new systems into any organisation. In concurrent engineering systems where departmental or discipline boundaries are crossed, job security is a serious concern. The introduction of these systems has major impacts on the company's organisation.

Each company has a different product profile and business strategy. The concurrent engineering organisation to employ the integrated CAD systems has to be customised for each company. The important issues for organisation and implementation are discussed in this chapter.

## 11.2 Organisation Impact

Organisation culture and methods represent the way a company work. They mould the mind set of the individuals in the company and have a direct effect on the person's willingness to communicate. The organisation issues for successful concurrent engineering are addressed in the following sections.

### 11.2.1 Product Development Process

Concurrency in the product development process requires process planning and factory design to begin before the product design is finalised. Partially completed assembly plans could be invalidated by changes in product design. The advantage of shortening the product lead time is balanced against the risk of confusion and inefficiency. Concurrent engineering requires the continuous exchange of information between members in the product development team. This exchange supports design decisions with a full understanding of the design on assembly and manufacture. The process engineers can clarify the constraints and opportunities for the product. Multidisciplinary teams are essential elements in concurrent engineering.

### 11.2.2 Multidisciplinary Teams

Multidisciplinary concurrent engineering teams are formed to bring together all the necessary expertise to complete the product. Effective concurrent engineering relies on challenging design opportunities with process constraints and alternatives. Only staff suitably empowered can positively contribute to this activity. The willingness of staff to take initiative in exchanging ideas and information and the authority to act on the agreement in team meetings is fundamental to success in any form of teamwork. The collocation of the team in the same workplace provides the environment for the informal exchange of information.

### 11.2.3 Role of Engineers

In the concurrent engineering environment, engineers perform their tasks with the additional input from colleagues in other disciplines. The integration of functions in product and process design broadens the scope of work of the engineers. They are members of the same integrated problem-solving team to achieve a better product. The engineers gain a cross-discipline knowledge and become more capable. They need team working and communication skills in addition to the specialist engineering knowledge. The role of the engineers is to achieve compromise with colleagues that results in a better product, not suboptimising their particular area of responsibility.

### 11.2.4 Project Management

The structure of projects governs the composition of project teams and the coordination of participation. Concurrent engineering team members rely on early sharing of information to begin their tasks. The formal specification of information release times and data packages is essential to the team working together. Appropriate project performance measures on timeliness, quality and cost on overall project achievement and interlinked stages can prevent suboptimisation of results. The degree of cohesion and trust between team members affects the flexibility that the project management has in redirecting resources to react in changes.

### 11.2.5 Management Structure

The team-based nature of concurrent engineering requires changes in the management structure. The reward a staff member receives in contributing to concurrent engineering must reflect the benefits to the organisation. In a functional organisation, the functional manager responsible for promotion and career development may not value a member's contribution to cross-functional efforts in performing work that benefits another department or another business. A structure to recognise and reward contribution that facilitates overall improvement must be implemented.

## 11.3 Implementation

The concurrent engineering culture is needed for the successful introduction of integrated systems. The communication of the reason for change to the staff, motivating and gaining their acceptance is of greater importance than the hardware and the software. All levels of staff must be involved, from senior management through to shopfloor personnel. They need to know how the changes will affect them personally. As the systems need to be populated with company specific rules, the people involved would be providing the key input. The implementation plans have human, technical and organisation aspects.

### 11.3.1 Target Organisation

The product development process of the implementing company needs to be mapped out and communicated to the staff. Introducing a major system that changes the way of work in the organisation is a good time to review operation methods. A business process re-engineering programme that formally maps out the structure of the activities in the organisation could lead to a leaner organisation. The new way of work must be communicated to the staff. This clear visibility will allay any fear of job security. The work model also illustrates the relationship and responsibility of the functional departments and individuals. The additional skills and resources needed can be identified in the re-engineering programme.

### 11.3.2 Senior Management Support

Strong and persistent senior management support is needed to sustain the change in organisation culture. There will be many who resist changes to their way of work and responsibility. The organisation changes will cut across the established power base and sphere of influence. The change implementation programme must be led by a senior manager who has access to the senior management. There should be regular review meetings for the progress to be monitored.

### 11.3.3 Awareness Programme

Prior to the introduction of the systems, all staff members must be made aware of what actually occurs and how the organisation will be affected. The aim is to eliminate the potential for mistrust and misunderstanding. The awareness must be given to all levels of staff with a suitably tailored programme for optimum communication. The importance to the business competitiveness and continued prosperity of the organisation must be communication. A multilayered plan must be developed. The purpose of the plan is to:

- inform;
- stimulate; and
- produce interaction.

It is essential to stimulate and produce interaction because the success of the system depends on the people in the organisation to become creative and contributive. The awareness programme must also include suppliers and customers because the boundary of the concurrent engineering integration extends out to all partners involve with the success of the product.

A whole range of media for consistent, caring communication is needed. This covers keynote briefings of the entire workforce to small group presentations with opportunities to answer questions. Newsletters, display boards and discussion group are other activities that occur in parallel. The awareness programme is a continuous process, not just something acting in short bursts. The success of pilot and ongoing projects need to be communicated so that other staff could be encouraged and excited. Multiple format and media should be used to extend the range of coverage. The basis of the awareness programme is:

$$involvement + enjoyment = ownership + retention.$$

A good awareness programme contributes to the breakdown of functional barriers, create more business awareness and team spirit.

## 11.3.4 System Customisation

Effective use of the integrated CAD system as a development method to support design decisions requires the customisation and setting up of the company's specific product structure and design rules. Current product models need to be analysed and standard modules and components need to be converted to the new system. Knowledge of product design, process planning and factory design needs to be elicited. These are compared with the system to identify the necessary work to prepare the system for use by the company. The customisation should be carefully documented as the basis for future systems maintenance and development.

## 11.3.5 Training

Training on the CAD system is needed for the product and process engineers. All members in the concurrent engineering team should be trained to use the integrated system. A feeling of equal access and ownership to the product and process models in the design system are essential for full and effective contribution from all members of the team.

Training on inter-personal skills, team working and group problem solving is also needed to develop the ability of the staff to work in concurrent engineering teams.

Real company-specific case studies should be developed for training where possible. The cost of training should be included as part of the continuous improvement investment in the staff.

## 11.3.6 Pilot Project

To establish the concurrent engineering way of work and refine the system implementation, a pilot project should be conducted. In order to ensure acceptance and integration, the pilot project must be:

- wanted and good;
- subject to real time constraints;
- adequately resourced and staffed; and
- subject to regular monitoring process.

The success of the pilot project illustrates the benefits of the concurrent engineering approach and convinces people of its significance. Any lessons from the pilot are used to adjust the rest of the implementation programme.

## 11.3.7 Monitoring and Auditing

Monitoring is the maintenance of a watch on the overall operational function of the concurrent engineering process. Productivity and quality improvements should be measured and maintained. This information provides valuable feedback for refining the system implementation. Concurrency in the product development process is achieved through the synergy between user and system and the removal of organisation boundaries. Failure to achieve the full benefits will occur if there is a lack of management commitment, too much is attempted too soon or the system is poorly maintained. The system audit will provided an environment that will help to prevent such problems.

## 11.4 Conclusions

The introduction of an integrated CAD system into a company affects a change of the organisation towards concurrent engineering practice. This has a significant impact to the organisation culture and structure. The implementation process needs to be carefully planned and engineered for maximum business benefit. Successful implementation requires continuous management effort and education of workforce. The implementation process is costly and time consuming. The alternative of not practising concurrent engineering reduces market share. Companies have no time to waste in developing an effective implementation plan and introducing the necessary system for concurrent engineering.

## 11.5 Bibliography

Ackoff, R. L. (1970) *A Concept of Corporate Planning*, Wiley, New York.

Adachi, T., Shih, L. C. and Enkawa, T. (1994) Strategy for supporting organisation and structuring of development teams in concurrent engineering, *Intl J Human Factors in Manufacturing*, **4**, 101–102.

Carter, D. E. and Baker, B. S. (1992) *Concurrent Engineering: The Product Development Environment for the 1990s*, Addison-Wesley, Reading, MA.

Clark, K. B. and Fujimoto, T. (1991) *Product Development Performance*, Harvard Business School.

Cross, N. (1994) *Engineering Design Methods: Strategies for Product Design*, 2nd Edn, Wiley, New York.

Ettlie, J. E. and Stoll, H. W. (1990) *Managing the Design–Manufacture Process*, McGraw-Hill, New York.

Galbraith, J. R. (1973) *Designing Complex Organisation*, Addison-Wesley, Reading, MA.

Institution of Electrical Engineers (1994) *IEE Colloquium on Issues of Cooperative Working in Concurrent Engineering*, June, London.

Institution of Mechanical Engineers (1994) Conference Papers from the *International Conference on Design for Competitive Advantage*, March, Coventry.

Joon-Hyo Kim (1994) *Applications of Design Conference to Support Cooperative Working in the Concurrent Engineering Environment*, MSc Thesis, The CIM Institute, Cranfield University.

McGrath, M. E., Anthony, M. T. and Sharpiro, A. R. (1992) *Product Development*, Butterworth-Heinemann, Oxford.

Kidd, P. T. and Karwowski, W. (eds) (1994) *Advances in Agile Manufacturing*, IOS Press.

Lettice, F. (1994) *Concurrent Engineering: A Team-Based Approach to Rapid Implementation*, Ph.D., Thesis, The CIM Institute, Cranfield University.

Parsaei, H. R. and Sullivan, W. G. (eds) (1993) *Concurrent Engineering*, Chapman and Hall, London.

Shina, S. G. (ed.) (1994) *Successful Implementation of Concurrent Engineering Products and Processes*, Van Nostrand Reinhold, New York.

Susman, G. I. (ed.) (1992) *Integrating Design and Manufacture for Competitive Advantage*, Oxford University Press.

Syan, C. S. and Menon, U. (1994) *Concurrent Engineering: Concepts, Implementation and Practice*, Chapman and Hall, London.

Wheelwright, S. C. and Clark, K. B. (1992) *Revolutionising Product Development—Quantum Leaps in Speed, Efficiency and Quality*, Macmillan, London.

## 11.6  Reference

Bostock Marketing Group (1990) *Market Survey "Managing Design in the 1990s"*.

# 12 *Conclusions*

This book aims at specifying computer-aided tools to be used in a concurrent engineering environment. The field addressed is the product, process, equipment and control systems design in the assembly domain.

Therefore, it proposes a functional architecture to address requirements of industrial users in both the off-line design and on-line manufacturing stages. The key point is the dual approach, which allows closure of the gap between the off-line and the on-line worlds.

The CAD method for industrial assembly provides a suite of tools that operate in a concurrent way by providing:

- the free exchange of data between modules that is possible throughout the design process and through to the operation of the workshop;
- the ability to use any one of the tools from very early within the design process;
- the complete integration of the suite of tools.

The *product design module* consists of five submodules each of which has different data requirements and rule structures. The user has total control of the sequence of analysis through the multiple starting points of the analysis. With an integrated database, the amount of repetitive user input is minimised. As much information as possible is extracted from the CAD solid model and from other integrated modules. The submodules are designed so that they can analyse a design throughout the design process.

The *assembly planner* has brought some advances in the field of assembly planning. Indeed, many features of this assembly planner (specification of an editor, concurrent engineering, . . .) are not often the aim of other ones.

The product model is well defined, including all features necessary to assembly planning: qualitative and functional attributes, rule-based technology module, etc. The concept of *actions* (i.e. the process model) is fully defined. An automatic scenario is also fully specified, containing precedence-constraint generation, assembly-tree generation and selection, and performance indices.

The *resource planner* constitutes a user-friendly tool for elaboration of high-quality logical layouts of assembly lines, i.e. it assigns operations to workstations along the line and selects equipments to carry out those operations, by minimising the *cost* of the line.

Even when working with incomplete data, i.e. in the context of concurrent engineering, the resource planner can provide useful information to the product/production means design team: an absence of equipments for a certain operation, or prohibitive price of any equipment capable of carrying out the operation, constitute a useful hint that the product should probably be redesigned using different assembly technologies. Such an advice can be usefully handled in particular by design for assembly. An impossibility to usefully combine together operations on workstations, leading to long idle times, constitutes a hint to the assembly planner that a different assembly sequence should be adopted.

Once the logical layout is elaborated, the physical layout editor is used to decide of the position of the workstations and their equipments on the shop floor. The trasportation devices (conveyors, AGVs, etc.) and possible buffers are designed as well. At that stage, a complete model of the future assembly line has been obtained, and the behaviour of the line can be *simulated* in order to assess the validity of the design.

The crucial part in any computer aided concurrent engineering system is to bridge the knowledge gap between the on- and off-line modules. The *simulation module* is used to integrate and validate the production planning modules as well as the control modules.

The user has the possibility to validate at a very early planning stage the off-line module decisions, such as for example the layout of the line or the dimensioning of buffers. Concurrently he or she is able to supply the flow control with a simulated system in order to test and validate the flow control rules.

The *scheduling module* schedules customer demands or production orders that were planned by the production planning system, creates bills of work for the resources that will process the orders and provides them for the flow-control module for order-release and down-loading. This module not only replies to industrial needs in general, but has applied them as an important guideline during its elaboration. These needs are nevertheless changing at a significant rate, due to changing markets and strategies, and as importantly, to new system supervision hardware and approaches. This module already satisfies most of these needs. However, if the size of the scenario is very big, the module would greatly profit from increased distributed architecture capabilities. These architectures are supported by new and innovative scheduling approaches, particularly in the field of artificial intelligence.

The scheduling module has most of the basic requirements to be compatible with a distributed architecture.

We have introduced new concepts in the *flow control module*. This module is closely related to the scheduling module. It performs the order release according to the chosen schedule. It also processes the real-time decisions that enable the flows of parts to move within the layout.

Finally, the book presents how to proceed with incomplete data in a concurrent engineering environment and how to introduce the new CAD method in industrial companies.

# Index

abstraction 43
additional action 28
additional process 104, 105, 108
adhesive bonding operation 104, 105, 113
AGV (*see* Automated Guided Vehicle)
architecture 10, 15, 24, 27–29, 35, 43, 46, 189, 190–201, 210, 213–215, 219
assembly automation 4
assembly constraint (*see* precedence constraint)
assembly line 4, 9, 12, 97, 100, 103, 107, 110, 113, 123, 129–133, 136, 139, 141, 143, 144, 149, 150, 155–158
assembly movement 104, 105, 106, 113, 114
assembly order 19
assembly plan 8, 10, 19, 23, 24, 27, 46, 129, 132, 136, 142, 145, 242, 243, 245, 246, 250–254, 261, 264
assembly plan editor 110, 125, 250
assembly plan evaluation 97, 99–101, 124
assembly plan selection 97, 99, 101, 118, 124
assembly plan visualization 114, 124
assembly planner 8, 22, 23, 27, 46, 61, 62, 64, 68, 76, 84, 144, 158, 241–243
assembly planning 4, 10, 14, 19, 22, 33, 41, 53, 95–127, 129, 238–240, 245, 250, 263

assembly process 3, 8, 18, 30, 32, 34, 36, 220, 223, 233, 234
assembly sequence 8, 23
assembly system 8, 19, 31, 40
assembly technology 100, 106, 115, 122
assembly tree 99, 110–115, 123–125
associativity 16, 237, 251
asynchronous polling 229, 235
attachment database 45
Automated Guided Vehicle 157, 158, 165, 171, 178
automated manufacturing 167, 170, 171
automated scheduling 11, 189, 192
automatic scenario 101, 106, 109, 125

base component 60–62, 87
base part 103, 108, 113
basic actions and resources 32
batch 4, 215, 219, 222, 228
batch transfer 205–209
bill of actions 106, 107, 110, 114
bill of constraints 106, 109, 110, 114
bill of material 31, 98, 103, 111–114
bill of work 34, 37, 169, 172, 174, 175, 177, 189, 197, 198, 202, 206, 208–210, 215, 218, 221, 229, 232
bin packing 134, 135
BOA (*see* bill of actions)
BOC (*see* bill of constraints)

BOM (*see* bill of material)
bottleneck 197, 210
BOW (*see* bill of work)
branch-and-bound 134, 153
breakdown 2, 161, 165, 166, 172, 174, 177, 178, 187
buffer 136, 158, 163, 165, 172, 174, 177, 178, 182–184, 187

case studies 4, 6
CE (*see* concurrent engineering)
cells (islands) 2, 129
centre of gravity 67, 70, 76
CIM architecture 37
cluster 98, 103, 113
combinatorial problem 98, 99, 103, 115, 120
communication module 48
company planning 35
compensation strategy 222, 233
component functionality 54, 55, 59–61, 64, 65
computer supported cooperative work 16
concurrent engineering 1, 2, 4, 7, 9, 13–22, 51, 57, 162, 168, 187, 217, 236, 263–269
constraint 189, 192, 195, 196, 198, 200, 202–213
control flow 10, 29, 35, 37, 42
control function 1, 9, 19, 20
control level 47
control module 12, 34, 36
control parameter 8, 9
control step 34
control system 3, 15, 16, 18, 27, 32, 36
conveyor 130–132, 136, 145, 157, 158, 164, 165, 171, 177–179
cost 3, 4, 11, 13, 14, 17, 23, 51, 54, 82, 86, 87, 265, 268
cost minimisation 8
cubic circuit 28, 40
customer demand 9, 36, 40, 172, 189, 197, 198, 200, 205
cycle time 11, 131–134, 148, 153, 154, 164, 165, 172, 174

data requirements 61, 64, 65, 72, 80, 84, 90, 92
data structure 17

decreasing priority rule 230
default rule 230
delivery 3, 30, 107, 111
design flow 10, 29, 33, 37
Design for Assembly 1, 2, 4, 10, 19, 22, 23, 33, 41, 45, 51, 56, 57, 92, 129, 158, 240, 241, 244, 245, 252
Design for Manufacture 15
design level 45
design process 9, 13, 17, 19, 20, 22–24, 51–53, 57, 92, 237, 240, 248, 262
design step 27, 33
destination rule 230
determinism 45, 47–49
deterministic system 40
DFA (*see* Design for Assembly)
dialectic aspect 27
discrete event simulation 166

earliest component 104, 122–124
ease of assembly 10, 19
element state 180, 181, 183, 184
epistemological approach 35, 40, 49
equipment database 8, 45, 66, 67, 74, 84–86, 136, 137, 141, 143, 148, 174
error recovery 11, 192, 198, 199, 205, 219
European Commission 15, 18
event 8, 12, 162, 167, 172–181, 183, 186, 187, 189, 190, 199, 205, 213, 218, 225, 229
execution control 36
execution level 48
execution step 34
exit 169, 172, 178, 179, 183
expanded cubic circuit model 42

factory design 20, 238, 240, 251, 263, 264, 267
fastener 98, 104, 108
fastening feature 63, 91
feeder 169, 172, 178, 179, 183
feeding 8, 29, 51–58, 61, 64, 71, 73–81, 84, 107, 111, 165, 178
FFD (*see* First Fit Descending)
finite loading 196, 201–203, 206, 209

# Index

First Fit Descending 134
flat surface 55, 66, 67, 71, 72, 74–76
flow control 1, 2, 8, 12, 19, 24, 161, 162, 164, 168, 169, 171–174, 176, 177, 215–235, 246, 248–250, 261, 262
flow control module 12, 34, 36, 46, 47
flow rule 230
flow shop 193
fractal model 31
functional 7
functional attribute 103, 125

geometrical data 8, 97, 98, 100–104, 106, 109, 115
geometrical link 104
geometrical precedence constraint 106, 109, 114
GGA (*see* Grouping Genetic Algorithm)
global decisions 169, 217, 218, 223
global flow control 9
global flow rules 169, 175, 177
global scheduling 11, 192, 193, 195, 196, 198, 200, 201, 203, 208, 210
graph search 97, 99, 114, 115, 122, 124
gripper 68–73
Grouping Genetic Algorithm 134, 135, 155

handling 8, 51, 53–58, 61, 63–68, 71–73, 76, 79, 80
handling feature 66–68, 72, 73
heuristic 97, 99
homomorphism 35
hyperarc 115–117, 119, 123

ideal architecture 29, 44
idle time 147, 151, 158
implementation 11, 12, 56, 263, 264, 266, 268, 269
indexing table 131, 151
infinite loading 196, 201–203, 206–210
input rule 230, 232
insertion 8, 54, 56, 58, 60, 81–86, 96, 98, 104, 111, 122

inspection 30, 40, 96, 107, 113
instantiation 43

jamming 165, 172, 174
jam 172, 177, 178, 187
job shop 129, 193, 210
job shop manufacturing 170
joining 8, 51, 53–58, 60, 61, 64, 85–87, 89–92
JIT (*see* Just In Time)
Just In Time 166, 192, 193, 215, 218, 233

knowledge level 44

latest component 104, 122–124
layout 1, 9, 11, 161, 163, 166, 168, 170, 172, 174, 178, 187
lead time 2, 14
level of reality 29, 32, 42
Line Balancing 8, 134, 151, 152, 155
linear line 131
link 162, 168, 174, 178, 183
local decisions 217, 222
local flow control 9
local flow rules 169, 175
local scheduling 11, 192, 193, 195–197, 199–201, 203, 213
logical layout 8, 11, 19, 27, 33, 34, 46, 133, 136, 142–144, 146, 155–158, 164, 171
loop 8, 132

macro-action 106, 108, 111, 125
man–machine interface 17, 23, 24, 198
manual assembly 163
manual scenario 101, 110
manufacturing 1, 8, 10, 11
manufacturing order 172
manufacturing processes 4, 8
marketing 2, 129, 131, 133
material flow 10, 28, 29, 35, 37, 42
material handling system 166
material transport 170, 178, 179, 185
micro-action 106, 108, 111
MMI (*see* man–machine interface)
model element 179, 180, 183, 185
monitoring module 8, 34

MRP 193, 199, 200
multicriteria approach 118
multidisciplinary teams 3, 14

obstacle database 109
off-line module 8, 19, 22–24
on-line module 8, 11, 24
OPT 210
order release 36, 47, 215
output rule 230
overlap transfer 203, 205–209

painting operation 107, 111, 113
pallet 164, 178, 179
parallel surface 71, 72
parallelism 96, 100, 122–125
part state rule 231
partial assembly tree 114, 115, 117–119, 123, 125
performance index 114, 115, 119, 120, 124
perturbation 40
Petri net 167, 181–187
physical layout 8, 11, 133, 136, 149, 156–158, 169, 178
physical layout editor 19, 157, 171, 174
PLC 9, 169, 217, 218, 225, 226, 229, 230, 233
polling request 177
pondered cyclic rule 230
position and insertion 52–54, 57, 81–83
positive advice 240
practical layout 34
precedence constraint 8, 9, 30, 32, 99, 101, 103–106, 109, 110, 114, 115, 117, 124, 125, 132, 134, 135, 142, 144–146, 148–150, 155
product cost 2, 14
product data 8, 18, 240
product data management 262
product design 2, 8, 10, 217
product development 2, 13, 23, 51, 238, 240, 264, 266, 268
product family 39
product model 9, 95, 97, 101, 102
product quality 14
product redesign 10

product structure 10, 54, 55, 57, 59, 267
production frame 29, 37
production interval 33
production mean 3, 4, 100, 123, 129, 133, 158
production order 9, 11, 35, 36, 40, 171, 189, 195, 197, 198, 200, 202–207, 209
production planning 35
production volume 4, 129, 131–133
prototyping 34
prototyping step 34

qualitative attribute 103, 104, 123

railway signalling 224, 225, 227
reactive scheduling 189, 194
real-time decision 9, 11, 215, 218, 233
real-time scheduling 11, 189, 195, 199, 200
redesign rule 2
reorientation 100, 113, 114, 119, 122, 124, 125
request rule 231–233, 235
rescheduling 192, 193, 198, 199, 205
research step 33
resource planner 2, 8, 9, 11, 22, 23, 27, 46, 85, 161, 169, 171, 174, 241–246
resource planning 2, 19, 32, 33, 41, 53, 95, 108, 125, 129–159, 164, 217, 240, 254
resource state 34
rough design 27, 46
routing 218, 221, 225, 232
routing decisions 221, 223
routing rule 215, 232, 233, 235
rules 53, 54, 61, 63–66, 68, 72, 74–76, 79, 80, 83, 84, 86, 90, 96, 97, 103, 106, 109, 115, 117, 119, 120, 266, 267

safety 224–226
scheduling 1, 2, 9, 11, 19, 24, 36, 44, 47, 164, 167, 171, 189–214, 246–248, 256, 261
scheduling method 11

# Index

scheduling module 8, 9, 11, 34, 46, 172, 215, 218, 219, 221
scheduling strategy 11, 161, 163, 168, 191, 194, 198, 202, 203, 210, 211
SCOPES 3, 4, 15, 18, 20, 53
screwing operation 98, 103, 105–108, 110, 111, 113
sectional release route locking 225, 228, 229
selection of equipment 11, 136, 155
set of certainties 28
set of hypotheses 27, 28
shop-floor control 1, 2, 20
simulation 1, 2, 9, 11, 16, 19, 24, 52, 53, 84, 85, 108, 158, 161–188, 217, 219, 224, 226, 230, 233, 245–251, 256, 257, 261, 263
simulation level 45, 46
simulation module 8, 9, 11
solid model 8, 10, 237
SPC 34, 222
SQC 34, 222, 228, 235
stability 100, 101, 122, 125
standard component or part 16, 55, 59, 61, 65, 87, 103–105, 108
subassembly 20, 96–98, 100, 101, 103, 115–117, 122–124
supervision system 8
supervisor 8, 9, 12, 34, 41, 47, 48, 191, 194, 195, 198, 204, 206, 210, 216–219, 222–225, 227, 229, 231–235
switch 162, 164, 169, 172, 175, 177, 183
switching element 216, 218, 222, 223, 226, 227, 229, 230
switching rule 33, 36, 162, 172, 216, 218, 229, 230, 232, 235
symmetry 67, 71, 73, 75, 77
system architecture 170, 172
system state 11, 34

teams 2, 3, 5, 13, 16–20, 51, 57, 59 263–265, 267, 268

technological data 95, 98
technological link 104
technological precedence constraint 10, 109, 110, 114
time horizon 37
time element 179
time-to-market 2, 3, 13, 14, 132
tolerance 56, 61–63, 66, 68, 72–76, 81–84
tolerancing 56
tooling cost 59
topology 193, 205
traffic 218, 225, 231, 235
traffic decisions 221, 222
traffic strategies 223
transfer 4, 29, 30, 31, 34
transport-box 178
transport-device 178
transport-network 178
transport-node 178
trolley 178

user-defined heuristic 194, 205, 212
user-defined rule 8, 230
user-defined precedence constraint 110
user-defined strategies 219, 223, 227
user group 4, 67
user interface 17
user requirement 86

variant 4, 34, 39, 215, 218–220, 234

worker 165, 170, 172, 174, 178, 179, 183, 185
worker qualification 165, 170
work in progress 167
workshop management 9, 11
workstation precedence graph 144–146, 149
WPG (*see* Workstation precedence graph)